LONDON MATHEMATICAL SOCIETY LECTURE NOT

Managing Editor: Professor J.W.S. Cassel
Department of Pure Mathematics and Mathe
16 Mill Lane, Cambridge CB2 1SB.

1. General cohomology theory and K-theor
4. Algebraic topology, J.F.ADAMS
5. Commutative algebra, J.T.KNIGHT
8. Integration and harmonic analysis on compact groups, R.E.EDWARDS
9. Elliptic functions and elliptic curves, P.DU VAL
10. Numerical ranges II, F.F.BONSALL & J.DUNCAN
11. New developments in topology, G.SEGAL (ed.)
12. Symposium on complex analysis, Canterbury, 1973, J.CLUNIE & W.K.HAYMAN (eds.)
13. Combinatorics: Proceedings of the British Combinatorial Conference 1973, T.P.McDONOUGH & V.C.MAVRON (eds.)
15. An introduction to topological groups, P.J.HIGGINS
16. Topics in finite groups, T.M.GAGEN
17. Differential germs and catastrophes, Th.BROCKER & L.LANDER
18. A geometric approach to homology theory, S.BUONCRISTIANO, C.P. ROURKE & B.J.SANDERSON
20. Sheaf theory, B.R.TENNISON
21. Automatic continuity of linear operators, A.M.SINCLAIR
23. Parallelisms of complete designs, P.J.CAMERON
24. The topology of Stiefel manifolds, I.M.JAMES
25. Lie groups and compact groups, J.F.PRICE
26. Transformation groups: Proceedings of the conference in the University of Newcastle-upon-Tyne, August 1976, C.KOSNIOWSKI
27. Skew field constructions, P.M.COHN
28. Brownian motion, Hardy spaces and bounded mean oscillations, K.E.PETERSEN
29. Pontryagin duality and the structure of locally compact Abelian groups, S.A.MORRIS
30. Interaction models, N.L.BIGGS
31. Continuous crossed products and type III von Neumann algebras, A.VAN DAELE
32. Uniform algebras and Jensen measures, T.W.GAMELIN
33. Permutation groups and combinatorial structures, N.L.BIGGS & A.T.WHITE
34. Representation theory of Lie groups, M.F. ATIYAH et al.
35. Trace ideals and their applications, B.SIMON
36. Homological group theory, C.T.C.WALL (ed.)
37. Partially ordered rings and semi-algebraic geometry, G.W.BRUMFIEL
38. Surveys in combinatorics, B.BOLLOBAS (ed.)
39. Affine sets and affine groups, D.G.NORTHCOTT
40. Introduction to Hp spaces, P.J.KOOSIS
41. Theory and applications of Hopf bifurcation, B.D.HASSARD, N.D.KAZARINOFF & Y-H.WAN
42. Topics in the theory of group presentations, D.L.JOHNSON
43. Graphs, codes and designs, P.J.CAMERON & J.H.VAN LINT
44. Z/2-homotopy theory, M.C.CRABB
45. Recursion theory: its generalisations and applications, F.R.DRAKE & S.S.WAINER (eds.)
46. p-adic analysis: a short course on recent work, N.KOBLITZ
47. Coding the Universe, A.BELLER, R.JENSEN & P.WELCH
48. Low-dimensional topology, R.BROWN & T.L.THICKSTUN (eds.)
49. Finite geometries and designs, P.CAMERON, J.W.P.Hirschfield & D.R.Hughes (eds.)

50. Commutator calculus and groups of homotopy classes, H.J.BAUES
51. Synthetic differential geometry, A.KOCK
52. Combinatorics, H.N.V.TEMPERLEY (ed.)
53. Singularity theory, V.I.ARNOLD
54. Markov processes and related problems of analysis, E.B.DYNKIN
55. Ordered permutation groups, A.M.W.GLASS
56. Journées arithmétiques 1980, J.V.ARMITAGE (ed.)
57. Techniques of geometric topology, R.A.FENN
58. Singularities of smooth functions and maps, J.MARTINET
59. Applicable differential geometry, M.CRAMPIN & F.A.E.PIRANI
60. Integrable systems, S.P.NOVIKOV et al.
61. The core model, A.DODD
62. Economics for mathematicians, J.W.S.CASSELS
63. Continuous semigroups in Banach algebras, A.M.SINCLAIR
64. Basic concepts of enriched category theory, G.M.KELLY
65. Several complex variables and complex manifolds I, M.J.FIELD
66. Several complex variables and complex manifolds II, M.J.FIELD
67. Classification problems in ergodic theory, W.PARRY & S.TUNCEL
68. Complex algebraic surfaces, A.BEAUVILLE
69. Representation theory, I.M.GELFAND et al.
70. Stochastic differential equations on manifolds, K.D.ELWORTHY
71. Groups - St Andrews 1981, C.M.CAMPBELL & E.F.ROBERTSON (eds.)
72. Commutative algebra: Durham 1981, R.Y.SHARP (ed.)
73. Riemann surfaces: a view towards several complex variables, A.T.HUCKLEBERRY
74. Symmetric designs: an algebraic approach, E.S.LANDER
75. New geometric splittings of classical knots (algebraic knots), L.SIEBENMANN & F.BONAHON
76. Linear differential operators, H.O.CORDES
77. Isolated singular points on complete intersections, E.J.N.LOOIJENGA
78. A primer on Riemann surfaces, A.F.BEARDON
79. Probability, statistics and analysis, J.F.C.KINGMAN & G.E.H.REUTER (eds.)
80. Introduction to the representation theory of compact and locally compact groups, A.ROBERT
81. Skew fields, P.K.DRAXL
82. Surveys in combinatorics: Invited papers for the ninth British Combinatorial Conference 1983, E.K.LLOYD (ed.)
83. Homogeneous structures on Riemannian manifolds, F.TRICERRI & L.VANHECKE
84. Finite group algebras and their modules, P.LANDROCK
85. Solitons, P.G.DRAZIN
86. Topological topics, I.M.JAMES (ed.)
87. Surveys in set theory, A.R.D.MATHIAS (ed.)
88. FPF ring theory, C.FAITH & S.PAGE
89. An F-space sampler, N.J.KALTON, N.T.PECK & J.W.ROBERTS
90. Polytopes and symmetry, S.A.ROBERTSON
91. Classgroups of group rings, M.J.TAYLOR
92. Simple Artinian rings, A.H.SCHOFIELD
93. Aspects of topology, I.M.JAMES & E.H.KRONHEIMER (eds.)
94. Representations of general linear groups, G.D.JAMES
95. Low dimensional topology 1982: Proceedings of the Sussex Conference, 2-6 August 1982, R.A.FENN (ed.)
96. Diophantine equations over function fields, R.C.MASON
97. Varieties of constructive mathematics, D.S.Bridges & F.RICHMAN
98. Localization in Noetherian rings, A.V.JATEGAONKAR
99. Methods of differential geometry in algebraic topology, M.KAROUBI & C.LERUSTE
100. Stopping time techniques for analysts and probabilists, L.EGGHE

London Mathematical Society Lecture Note Series: 89

An F-space sampler

N.J. KALTON
University of Missouri, Columbia

N.T. PECK
University of Illinois, Urbana

James W. ROBERTS
Unversity of South Carolina, Columbia

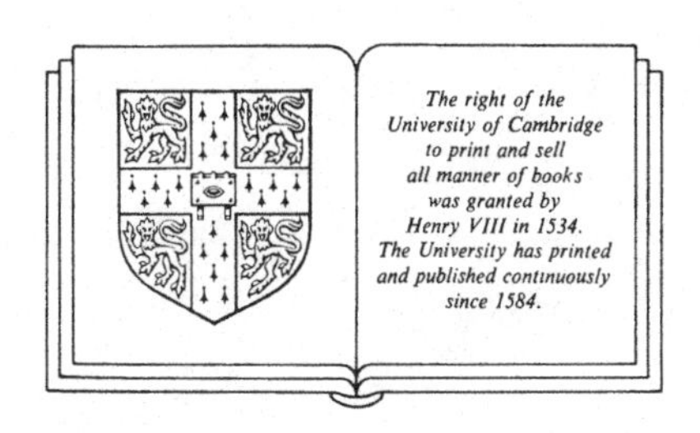

CAMBRIDGE UNIVERSITY PRESS
Cambridge
London New York New Rochelle
Melbourne Sydney

CAMBRIDGE UNIVERSITY PRESS
Cambridge, New York, Melbourne, Madrid, Cape Town, Singapore, São Paulo

Cambridge University Press
The Edinburgh Building, Cambridge CB2 8RU, UK

Published in the United States of America by Cambridge University Press, New York

www.cambridge.org
Information on this title: www.cambridge.org/9780521275859

First published 1984
Re-issued in this digitally printed version 2008

A catalogue record for this publication is available from the British Library

Library of Congress Catalogue Card Number: 84–45434

ISBN 978-0-521-27585-9 paperback

To Adrian and Verona Roberts

To Jennifer

To Emily

CONTENTS

PREFACE

Typically, a basic text on functional analysis will only make the briefest of references to general topological vector spaces, before restricting attention to the locally convex case or to Banach spaces. Thus most analysts are aware of the existence of non-locally convex spaces such as $L_p(0,1)$ for $0 < p < 1$ but know very little about them. The neglect of non-locally convex spaces is easily understood. The basic theory of Banach spaces, which sits at the core of modern functional analysis, may be said to depend on two major principles - the Hahn-Banach theorem and the Closed Graph theorem (which may be taken to include weaker theorems such as the Uniform Boundedness Principle). Working in non-locally convex spaces, even when they are complete and metrizable, requires doing without the Hahn-Banach theorem. The role of the Hahn-Banach theorem may be said to be that of a universal simplifier - infinite-dimensional arguments can be reduced to the scalar case by the use of the ubiquitous linear functional. Thus the problem with non-locally convex spaces is that of "getting off the ground." This difficulty in even making the simplest initial steps has led some to regard non-locally convex spaces as simply uninteresting. It's our contention, which we hope to justify in these notes, that this attitude is mistaken and that with the aid of fresh techniques one can develop a rich and fulfilling theory.

Our aim, therefore, in these notes is to present some aspects of the theory of F-spaces (complete metric linear spaces) which we hope the reader will find attractive. We do not aim to be encyclopaedic, nor do we strive for complete generality in the results which we present. The account is intended to be reasonably self-contained, at least for the reader versed in the basics of topological vector space theory (see e.g. Rudin [1973] or Kothe [1969]). For more background one should refer to Rolewicz [1972] which gives a fairly complete summary of the state of the art up to 1972.

In selecting the material for these notes we have adopted the theme of taking certain familiar properties or theorems from Banach space theory and examining their behavior in general F-spaces. Thus we shall consider in detail the fate of the Hahn-Banach theorem and the Krein-Milman theorem. Also the study of compact operators in a non-locally convex setting takes on a new twist, largely because much of the Fredholm theory can be extended. We place special emphasis on the classical examples of non-locally convex F-spaces - the sequence spaces ℓ_p $(0 < p < 1)$, the function spaces L_p $(0 \leq p < 1)$ and the Hardy spaces H^p $(0 < p < 1)$.

We now turn to a summary of the contents chapter by chapter.

<u>Chapter 1:</u> <u>Preliminaries</u>. This chapter, after the introduction, recalls some of the basic properties of F-spaces. Many readers will be familiar with the contents of this chapter, but a brief perusal is advisable if only to establish notation. We treat the problem of determining an invariant metric on a metric linear space from a slightly unusual point of view by introducing the idea of a Δ-norm. The Closed Graph and Open Mapping theorems are given in a particular form, which is perhaps not as well-known as it should be, since this form is required later in the book.

<u>Chapter 2:</u> <u>Some of the classical examples</u>. We study here the basic properties of the sequence spaces ℓ_p $(0 < p < 1)$ and the function spaces L_p for $0 \leq p < 1$. Part of the aim of the chapter is to familarize the reader with some of the techniques that can be used in studying F-spaces. We also show as motivation for Chapter 4 that the Hahn-Banach theorem fails in

each space (i.e. there is a continuous linear functional defined on a linear subspace which cannot be extended continuously to the whole space). For the spaces L_p, this amounts to showing that they have trivial dual spaces, i.e. every continuous linear functional is zero. However the sequence spaces ℓ_p have rich dual spaces, which we calculate; therefore to demonstrate the failure of the Hahn-Banach theorem requires rather more effort. In particular, we show that ℓ_p has a quotient with trivial dual and this implies the failure of the Hahn-Banach theorem.

Chapter 3: The Hardy Spaces H^p. The Hardy spaces H^p for $0 < p < 1$ provide another rich store of examples and have played an influential role in determining the general direction taken by F-space theory. We present here their more fundamental properties. In particular we calculate their "Banach envelopes" and demonstrate again the failure of the Hahn-Banach theorem. The spaces H^p have, like ℓ_p, rich dual spaces but again we show that they have quotients with trivial dual. As part of the chapter we present an elegant result of Aleksandrov that $H^p + \overline{H^p} = L_p$.

Chapter 4: The Hahn-Banach Extension Property. Motivated by the results of Chapters 2 and 3, we show that an F-space in which the Hahn-Banach theorem holds is locally convex. Curiously this result is false without the assumption of metrizability and it is unknown whether completeness is necessary. A separable non-locally convex F-space which has a dual rich enough to separate points always has a quotient space with trivial dual.

Chapter 5: The three-space problem. A closed subspace N of an F-space X is said to have the Hahn-Banach Extension Property (HBEP) if every continuous linear functional on N can be extended to a continuous linear functional on X. The main result of Chapter 4 could be restated as saying that every non-locally convex F-space has a closed subspace which fails HBEP.

In Chapter 5 we study conditions on the quotient space X/N for a closed subspace N which imply that N has HBEP; a space Y is called a K-space if $X/N \approx Y$ implies N has HBEP. The main results of the chapter are that the spaces ℓ_p, L_p $(0 < p < \infty, \neq 1)$ are K-spaces. However an example due to Ribe is constructed to show that ℓ_1 is not a K-space; precisely, the Ribe space is a non-locally convex quasi-Banach space X which has a subspace N of dimension one so that $X/N \approx \ell_1$. The Ribe space is also an example of a separable F-space with no (non-trivial) quotient space with trivial dual.

Chapter 6: Lifting theorems. This short chapter reformulates some of the ideas of the previous chapter as lifting theorems for operators. We prove some general lifting theorems for L_p $(0 \leq p < 1)$ and deduce that L_p is a K-space.

Chapter 7: Transitivity and small operators. In Chapter 7 we broaden our interests from linear functionals to general compact operators. The general theme of the chapter is that it is quite difficult to find compact operators on spaces which do not already admit linear functionals. To make this notion precise we introduce the important idea of transitivity. X is transitive if for every $x_1, x_2 \in X$ with $x_1 \neq 0$ there is an endomorphism T of X so that $Tx_1 = x_2$. We show that on a transitive space with trivial dual there are no non-zero strictly singular (and hence no compact) endomorphisms; the reason is essentially that the Fredholm theorem can be extended to this setting (with some modifications). The spaces L_p $(0 < p < 1)$ are transitive, but for these spaces we prove stronger results. Any non-zero operator $T: L_p \to X$ (for any range space) is an isomorphism on an infinite-dimensional Hilbertian subspace of L_p.

To put these results in perspective we also construct a space with trivial dual which does have non-zero compact endomorphisms. This space is of course not transitive. A more spectacular non-transitive space is the rigid space constructed at the end of the chapter: this is a quasi-Banach space whose algebra of endomorphisms consists simply of multiples of the identity operator.

<u>Chapter 8: Operators on</u> L_p, $0 \leq p < 1$. The main theorems of this chapter give an explicit representation for all the endomorphisms of L_p for $0 \leq p < 1$. Using these theorems we can extend the results of Chapter 7. We see that every non-zero endomorphism of L_p $(0 \leq p < 1)$ actually preserves a copy of the whole space. We also sketch a proof that every non-zero operator $T : L_p \to L_0$ for $0 < p < 1$ preserves a copy of any L_q for $p < q \leq 2$.

<u>Chapter 9: A compact convex set with no extreme points</u>. The final chapter is devoted to a distinct but not unrelated question: can one construct in a non-locally convex F-space a compact convex set which fails to have extreme points? The answer is yes and we construct an example in L_0. The crucial notion is that of a <u>needle-point</u>, which we first met in disguised form in Chapter 5 (the Ribe space).

At this point it seems appropriate to make some apologies. Some readers will feel that certain topics should have been (or should not have been!) covered - obviously the selection of the material in a volume of limited length has to be somewhat personal. To the authors, it seems that one obvious topic which might have been included is the Maurey-Nikishin factorization theory for operators into L_0 (briefly mentioned in Chapter 8). However, we wished to keep as closely as possible to certain predetermined themes.

Although these notes have three distinct authors, we have tried to be as consistent as possible. There are inconsistencies in style and notation, which we hope will not unduly distract the reader. We all wish to apologize on behalf of each other for any illiteracy or obscurity!

Finally, it is a pleasure to thank: Allen Butler, Raouf Eldeeb, Larry Riddle, Jon Snader, David Trautman, who participated in courses or seminars on this material and contributed useful ideas; Patricia Coombs, for her expert typing of the manuscript; David Tranah, of the Cambridge University Press, for his advice and assistance; and the National Science Foundation, for its support.

Columbia, MO
Urbana, IL
Columbia, SC
March, 1984

CHAPTER I
PRELIMINARIES

1. Topological vector spaces

A topological vector space X over a field $\mathbb{K}$ (either the real numbers $\mathbb{R}$ or the complex numbers $\mathbb{C}$) is a vector space such that the operations of addition $(x,y) \to x + y$ $(X \times X \to X)$ and scalar multiplication $(\lambda,x) \to \lambda x$ $(\mathbb{K} \times X \to X)$ are jointly continuous. We shall assume that the reader is familiar with the basic properties of topological vector spaces, as expounded for example in Kothe [1969] or Rudin [1973] Chapter 3. In the introduction we present selected basic material which will be of special interest to us, omitting most of the routine proofs.

We recall that a vector topology τ on X can be described in terms of a base of neighborhoods $\mathcal{U}$ at the origin 0. A base $\mathcal{U}$ has the properties

(1) Given $U \in \mathcal{U}$ there exists $V \in \mathcal{U}$ with $V + V = \{v_1 + v_2 : v_1, v_2 \in V\} \subset U$.

(2) Given $U \in \mathcal{U}$, there exists $V \in \mathcal{U}$, with $\alpha V \subset U$ for all $\alpha \in \mathbb{K}$ with $|\alpha| \leqslant 1$.

(3) Given $U \in \mathcal{U}$ and $x \in X$ there exists $n \in \mathbb{N}$ with $x \in nU$ (i.e. U absorbs $\{x\}$).

Conversely given a collection of sets $\mathcal{U}$ obeying (1)-(3), there is a vector topology τ on X with $\mathcal{U}$ as a neighborhood-base for the origin.

It is always possible to choose a base of neighborhoods of sets which are closed and balanced, i.e. $\lambda U \subset U$ for $|\lambda| \leqslant 1$.

(X,τ) is Hausdorff if and only if $\cap\, \mathcal{U} = \{0\}$. In general $\cap\, \mathcal{U}$ is a linear subspace of X. (X,τ) is locally convex if it has a base of neighborhoods $\mathcal{U}$ consisting of convex sets ; a set C is convex if $\lambda C + (1-\lambda)C \subset C$ for

$0 \leqslant \lambda \leqslant 1$. It is then possible to choose a base of <u>absolutely convex sets</u> U (i.e. such that $\lambda U + \mu U \subset U$ for $|\lambda| + |\mu| \leqslant 1$).

If M is a linear subspace of X then the quotient space X/M may be topologized by the quotient topology whose neighborhood base consists of all sets $(q(U) : U \in \mathcal{U})$ where $q : X \to X/M$ is the quotient map and $\mathcal{U}$ is a neighborhood base in X. X/M will be Hausdorff provided M is closed.

2. <u>Metric linear spaces</u>

We start with a rather non-standard definition which appears to be useful. Let X be a vector space. Then a Δ-<u>norm</u> is a map $x \to \|x\|$ $(X \to \mathbb{R})$ so that

(1.1) $\quad \|x\| > 0 \qquad x \neq 0$

(1.2) $\quad \|\alpha x\| \leqslant \|x\| \qquad |\alpha| \leqslant 1,\ x \in X$

(1.3) $\quad \lim_{\alpha \to 0} \|\alpha x\| = 0 \qquad x \in X$

(1.4) $\quad \|x+y\| \leqslant C \max(\|x\|, \|y\|) \qquad x, y \in X$

where C is some constant independent of x, y. Note that $C > 1$.

If $\|\cdot\|$ is a Δ-norm on X then it induces on X a vector topology τ which is metrizable. A base of neighborhoods at the origin is given by sets of the form $U_n = \{x \in X : \|x\| < 1/n\}$. A sequence $x_n \in X$ converges to $x \in X$ if and only if $\|x - x_n\| \to 0$.

Conversely suppose τ is a topology with a countable base of neighborhoods (U_n) such that $\cap U_n = \{0\}$, each U_n is balanced and $U_{n+1} + U_{n+1} \subset U_n$ for every n. Then we can define a Δ-norm on X by

$$\|x\| = \sup(2^{-n} : x \notin U_n)$$

and the Δ-norm induces the original topology; here $C = 2$.

A Δ-norm is called an F-<u>norm</u> if it satisfies

$$(1.5) \qquad \|x+y\| \le \|x\| + \|y\| \qquad x,y \in X.$$

If $\|\cdot\|$ is any F-norm then $d(x,y) = \|x-y\|$ is a (translation-)<u>invariant</u> <u>metric</u> on X. We first prove a metrization theorem which allows us to replace any Δ-norm with an F-norm.

LEMMA 1.1. Let $\|\cdot\|$ be any Δ-norm on X. Choose p so that $2^{1/p} = C$. Then for any $x_1,\ldots,x_n \in X$ we have

$$(1.6) \qquad \|x_1 + \ldots + x_n\| \le 4^{1/p}(\|x_1\|^p + \ldots + \|x_n\|^p)^{1/p}.$$

<u>Proof</u>. By induction on (1.4) we can obtain

$$(1.7) \qquad \|x_1 + \ldots + x_n\| \le \max_{1\le k\le n} C^k\|x_k\|.$$

for $x_1,\ldots x_n \in X$.

Let us define a function $H : X \to \mathbb{R}$ by

$$H(x) = 2^{n/p} \quad \text{if} \quad 2^{n-1/p} < \|x\| \le 2^{n/p} \qquad n \in \mathbb{Z}$$

$$H(0) = 0.$$

Then

$$\|x\| \le H(x) \le 2^{1/p}\|x\|.$$

We shall show by induction that

$$(1.8) \qquad \|x_1 + \ldots + x_n\| \le 2^{1/p}(H(x_1)^p + \ldots + H(x_n)^p)^{1/p}$$

and then (1.6) is immediate.

Of course (1.8) holds if $n = 1$. Suppose that it holds for $n = m$ and that $x_1, \ldots, x_{m+1} \in X$. We may suppose that $\|x_1\| \geq \|x_2\| \geq \ldots \geq \|x_{m+1}\|$. We consider two cases.

First suppose that the set of values $\{ H(x_i) : 1 \leq i \leq m+1\}$ is distinct: then $H(x_i) \leq 2^{1-i/p} H(x_1)$ and so

$$C^{i} \|x_i\| \leq C^{i} H(x_i)$$

$$\leq 2^{1/p} H(x_1)$$

$$\leq 2^{1/p} (H(x_1)^p + \ldots + H(x_n)^p)^{1/p}$$

and (1.8) follows from (1.7).

Alternatively we have $H(x_j) = H(x_{j+1})$ for some j, $1 \leq j \leq m$. Hence, for some $r \in \mathbb{Z}$,

$$2^{(r-1)/p} < \|x_{j+1}\| \leq \|x_j\| \leq 2^{r/p}$$

and so

$$\|x_j + x_{j+1}\| \leq 2^{(r+1)/p}.$$

Thus $H(x_j + x_{j+1})^p \leq H(x_j)^p + H(x_{j+1})^p$. Applying the inductive hypothesis,

$$\|x_1 + \ldots + x_{m+1}\|^p \leq 2(\sum_{i \neq j, j+1} H(x_i)^p + H(x_j + x_{j+1})^p)$$

and hence (1.8) follows for $n = m + 1$. □

THEOREM 1.2. Let $\|\cdot\|$ be a Δ-norm on X. Then if p is chosen so that $2^{1/p} = C$, the formula

$$(1.9) \qquad |||x||| = \inf\left(\sum_{i=1}^{n} \|x_i\|^p : \sum x_i = x\right)$$

defines an F-norm on X giving the same topology.

Proof. Simply note that $(1/4)\|x\|^p \leq |||x||| \leq \|x\|^p$. □

COROLLARY. Let X be a Hausdorff topological vector space with a countable base of neighborhoods of 0. Then X is metrizable and the topology may be given by an invariant metric.

A metrizable topological vector space (or metric linear space) is called an F-_space_ if it is complete for an invariant metric (and hence for every invariant metric.) An elegant result of Klee [1952] allows one to replace an invariant metric by _any_ metric giving the topology. Thus a metric linear space which is complete for any metric will also be complete in any invariant metric.

Every metrizable topological vector space X can be embedded as a dense linear subspace of an F-space $\tilde{X}$. The construction of $\tilde{X}$ is simply to form the normal metric space completion of X with respect to an invariant metric and extend the vector space operations in the obvious way. The space $\tilde{X}$ obtained in this way is unique; it does not depend on the particular choice of invariant metric.

If N is a closed subspace of a metrizable topological vector space then X/N is also metrizable. Further if X is an F-space, X/N is an F-space. Again we shall not prove these statements which are left as an exercise.

3. Locally bounded spaces

A subset B of a (Hausdorff) topological vector space X is bounded if given any zero-neighborhood U we have $B \subset nU$ for some integer n. Note that if $B_1, \ldots, B_m$ are bounded sets then so is $B_1 + \ldots + B_m$.

Suppose X has a neighborhood of the origin B which is bounded. Then the sets $(1/n\, B : n \in \mathbb{N})$ form a neighborhood-base at the origin. Hence X is metrizable. Also $B+B$ is bounded and so for some $\rho \geq 1$, $B + B \subset \rho B$. Define a Δ-norm on X by using the Minkowski-function of B:

$$\|x\| = \inf\{\lambda : \lambda^{-1} x \in B\}. \tag{1.10}$$

To see this is a Δ-norm note that

$$\|x+y\| \leq \rho \max(\|x\|, \|y\|). \tag{1.11}$$

In addition $\|\cdot\|$ satisfies

$$\|\alpha x\| = |\alpha| \|x\| \qquad \alpha \in \mathbb{K},\ x \in X. \tag{1.12}$$

A Δ-norm which satisfies (1.12) is called a quasi-norm. Conversely the topology associated with any quasi-norm is locally bounded. A complete locally bounded space is called a quasi-Banach space.

For $0 < p \leq 1$ we say that a subset C of a vector space X is p-convex if whenever $x_1, x_2 \in C$ and $a_1, a_2 \in \mathbb{R}$ with $a_1 \geq 0$, $a_2 \geq 0$, $a_1^p + a_2^p = 1$, then $a_1x_1 + a_2x_2 \in C$. C is absolutely p-convex if $a_1x_1 + a_2x_2 \in C$ whenever $x_1, x_2 \in C$, $a_1, a_2 \in \mathbb{K}$ and $|a_1|^p + |a_2|^p \leq 1$.

Suppose now that a locally bounded space X has a bounded absolutely p-convex neighborhood B of the origin.

Then the quasi-norm given by (1.10) satisfies

(1.13) $$\|x+y\|^p \leq \|x\|^p + \|y\|^p \qquad x,y \in X.$$

Conversely, if a quasi-norm $\|\cdot\|$ satisfies (1.13) then its unit ball $U = \{x : \|x\| \leq 1\}$ is absolutely p-convex. We say that the quasi-norm is then p-<u>subadditive</u> and the locally bounded space X is <u>locally</u> p-<u>convex</u> or merely p-<u>convex</u>. It is common to abbreviate "p-convex quasi-Banach space" to "p-<u>Banach space</u>," and it will be assumed, when this term is used, that the associated quasi-norm satisfies (1.13).

If we specialize Theorem 1.2 to this situation we have a classical result due to Aoki [1942] and Rolewicz [1957].

THEOREM 1.3. (Aoki-Rolewicz theorem). If X is a locally bounded space then X is p-convex for some $p > 0$. Precisely, if B is a bounded neighborhood of the origin with $B + B \subset \rho B$ then X is p-convex where $2^{1/p} = \rho$.

<u>Proof</u>. It follows from Theorem 1.2 that if

$$\|x\|_1 = \inf\left(\left(\sum_{i=1}^{n} \|x_i\|^p\right)^{1/p} : \sum x_i = x\right)$$

then $\|x\|_1$ is a p-subadditive quasi-norm equivalent to the given quasi-norm. □

REMARKS. The constant ρ in Theorem 4.3 is such that

(1.14) $$\|x+y\| \leq \rho \max(\|x\|,\|y\|) \qquad x,y \in X.$$

However it is more common to associate to the quasi-norm a constant k, the <u>modulus of concavity</u> of the quasi-norm, so that

(1.15) $$\|x+y\| \leq k(\|x\|+\|y\|) \qquad x,y \in X.$$

Of course, (1.14) implies (1.15) with $k = \rho$ while (1.15) implies (1.14) with $\rho = 2k$.

However Theorem 1.3 yields the fact that if (1.14) holds then X may be equivalently quasi-normed by a p-subadditive quasi-norm $\| \ \|_1$ and then

$$\|x+y\|_1 \leq (\|x\|_1^p + \|y\|_1^p)^{1/p}$$

which implies by elementary calculus

$$\begin{aligned}\|x+y\|_1 &\leq 2^{1/p-1}(\|x\|_1 + \|y\|_1) \\ &= 1/2\ \rho(\|x\|_1 + \|y\|_1).\end{aligned}$$

Thus by renorming we can take $k = 1/2\ \rho$. Note also that from Theorem 1.3 if (1.15) holds then X is p-convex where $2^{1/p} = 2k$ or $k = 2^{1/p-1}$.

4. Linear operators and the closed graph theorem

If X and Y are topological vector spaces (over the same base field $\mathbb{K}$), then a linear map $T : X \to Y$ will be called an <u>operator</u> when it is continuous. This is equivalent to the statement that for every zero-neighborhood U in Y, $T^{-1}(U)$ is a zero-neighborhood in X. When X and Y are metrizable and their topologies are given by Δ-norms (both denoted by $\|\cdot\|$) then T is continuous if given $\epsilon > 0$ there exists $\delta(\epsilon) > 0$ so that $\|x\| < \delta$ implies $\|Tx\| < \epsilon$. If X and Y are both locally bounded and have their topologies given by quasi-norms we need only have

$$\|T\| = \sup_{\|x\| \leq 1} \|Tx\| < \infty.$$

We denote by $\mathcal{L}(X,Y)$ the vector space of all linear operators $T : X \to Y$. If $Y = \mathbb{K}$, the base field, we call an operator a <u>linear functional</u> and write $\mathcal{L}(X,\mathbb{K}) = X^*$. If $Y = X$

then an operator is an endomorphism and $L(X,X)$ is abbreviated to $L(X)$.

In this section we shall prove the familiar Open Mapping and Closed Graph Theorems for F-spaces. Although these are to be found in any basic functional analysis book, we shall need certain strengthened forms which are perhaps less well known.

THEOREM 1.4. (The Open Mapping Theorem). Let X be an F-space and let Y be a Hausdorff topological vector space. Let $T : X \to Y$ be a linear operator such that for every zero-neighborhood U in X, $\overline{T(U)}$ is a neighborhood of zero in Y. Then

(1) T is an open mapping i.e. $T(U)$ is a neighborhood of 0 for every neighborhood U of zero in X,

(2) Y is an F-space.

Proof. (Essentially as in Rudin [1973], p. 47). Suppose X is F-normed by $\|\cdot\|$. Let

$$V(\epsilon) = \{x \in X : \|x\| < \epsilon\}.$$

We show $T[V(\epsilon)] \supset \overline{T[V(1/2\ \epsilon)]}$. Indeed suppose $y = y_1 \in \overline{T[V(1/2\ \epsilon)]}$. By induction we pick $y_n \in \overline{T[V(2^{-n}\epsilon)]}$, and $x_n \in V(2^{-n}\epsilon)$. Suppose y_n has been chosen. Then

$$(y_n + \overline{T[V(2^{-n-1}\epsilon]}) \cap T[V(2^{-n}\epsilon] \neq \varphi.$$

Pick $x_n \in V(2^{-n}\epsilon)$ so that

$$y_n - Tx_n \in \overline{T[\cdot\ 2^{-n-1}\epsilon)]}$$

and let $y_{n+1} = y_n - Tx_n$.

Now $\|x_n\| < 2^{-n}\epsilon$ and so

$$x = \sum_{n=1}^{\infty} x_n$$

exists in X and

$$\|x\| < \epsilon.$$

T is continuous, so that

$$\begin{aligned} Tx &= \sum_{n=1}^{\infty} Tx_n \\ &= \lim_{N\to\infty} \sum_{n=1}^{N} Tx_n \\ &= \lim_{N\to\infty} \sum_{n=1}^{N} (y_n - y_{n+1}) \\ &= \lim_{N\to\infty} (y_1 - y_{N+1}). \end{aligned}$$

However $y_{N+1} \in \overline{T[V(2^{-N-1}\epsilon)]}$ and so as T is continuous $y_{N+1} \to 0$. Thus $Tx = y_1 = y \in T[V(\epsilon)]$. This immediately establishes (1). Now it follows that $T(V(2^{-n}))$ is a base of neighborhoods in Y so that Y is isomorphic to the quotient X/ker T (where ker T = {x : Tx = 0}). Hence Y is an F-space. □

COROLLARY 1.5. Let X and Y be F-spaces and let $T : X \to Y$ be a surjective linear operator. Then T is an open mapping.

Proof. If U is a neighborhood of 0 in X, then choose another balanced neighborhood V with $V + V \subset U$. Then $\cup\, nT(V) = Y$ and so by the Baire Category Theorem, for some k int $\overline{kT(V)}$ is non-empty. Thus int $\overline{T(V)}$ is non-empty and so 0 is in the interior of $\overline{T(V)} - \overline{T(V)} \subset \overline{T(U)}$. Now apply Theorem 1.4. □

THEOREM 1.6. (The Closed Graph Theorem). Let X be a Hausdorff topological vector space and let Y be an F-space. Let

$T : X \to Y$ be a linear map whose graph $G(T) = \{(x,Tx) \subset X \times Y\}$ is closed. Suppose also that if V is a neighborhood of 0 in Y then $T^{-1}(V)$ is a neighborhood of 0 in X. Then T is continuous.

<u>Proof</u>. Let $\mathcal{U}(X)$ and $\mathcal{U}(Y)$ be bases of balanced neighborhoods in X and Y respectively. We define a new vector topology β on Y by specifying a base of sets of the form $\{T(U) + V : U \in \mathcal{U}(X), V \in \mathcal{U}(Y)\}$.

We claim first that β is Hausdorff. Indeed if $y_0 \in \cap [T(U)+V) : U \in \mathcal{U}(X), V \in \mathcal{U}(Y)]$ then for every $U \in \mathcal{U}(X)$, $V \in \mathcal{U}(Y)$ there exist $x \in U$, $y \in V$ with $y_0 = Tx + y$. Hence in $X \times Y$, $(0,y_0) = (x,Tx) + (-x,y)$ so that $(0,y_0) \in \overline{G(T)}$. Thus $y_0 = 0$.

Now consider the identity map $i : Y \to (Y,\beta)$. If $V_0 \in \mathcal{U}(Y)$ we claim that the β-closure of V_0 contains a neighborhood of 0 in (Y,β). Choose $V_1 \in \mathcal{U}(Y)$ with $V_1 + V_1 \subset V_0$. Then for some $U_1 \in \mathcal{U}(X)$,

$$U_1 \subset \overline{T^{-1}(V_1)} \subset T^{-1}(V_1) + U$$

for every $U \in \mathcal{U}(X)$. Thus

$$T(U_1) \subset V_1 + T(U) \qquad U \in \mathcal{U}(X)$$

so that

$$V_1 + T(U_1) \subset V_0 + T(U) \qquad U \in \mathcal{U}(X).$$

Hence $V_1 + T(U_1) \subset \overline{V_0}^{\beta}$.

Now by the Open Mapping Theorem, i is open and hence i^{-1} is continuous. Thus β is the F-space topology of Y and so if $V \in \mathcal{U}(Y)$ there exists $U \in \mathcal{U}(X)$ with $T(U) \subset V$ i.e. T is continuous. □

COROLLARY 1.7. Suppose X and Y are F-spaces and $T : X \to Y$ is a linear map with closed graph. Then T is continuous.

<u>Proof</u>. See the Corollary to Theorem 1.4.

REMARKS. The Closed Graph and Open Mapping Theorems in the form given above are essentially due to Pettis [1950]. See also Kelley [1963] p. 213.

5. <u>Bases and basic sequences</u>

A <u>basis</u> of an F-space X is a sequence (e_n) so that every $x \in X$ has a <u>unique</u> expansion in the form

$$x = \sum_{n=1}^{\infty} \alpha_n e_n .$$

For any basis (e_n) we can define partial-sum maps $S_n : X \to X$ by

$$S_n x = \sum_{i=1}^{n} \alpha_i e_i$$

and linear maps $e_n^* : X \to K$ by

$$e_n^*(x) = \alpha_n .$$

It is a nice classical application of the Closed Graph Theorem that these maps are in fact continuous (Banach [1932]).

THEOREM 1.8. If (e_n) is a basis of X then the linear maps e_n^* are continuous linear functionals on X and the partial sum maps (S_n) form an equicontinuous family of linear operators.

Proof. Let $\|\cdot\|$ be any F-norm on X. For $x \in X$ define

$$|||x||| = \sup_n \|S_n x\|.$$

Then $|||\cdot|||$ is an F-norm on X and $|||x||| \geq \|x\|$. To check that $|||\cdot|||$ is an F-norm requires one non-obvious step - we need

$$\lim_{\alpha \to 0} |||\alpha x||| = 0 \qquad x \in X.$$

In fact the sequence $\{S_n x : n = 1,2,\ldots\}$ is convergent and therefore bounded in X and this implies the desired property.

Next we note that $(X, |||\cdot|||)$ is complete. If (u_n) is a $|||\cdot|||$-Cauchy sequence, then for some $u \in X$, $\|u_n - u\| \to 0$. However $(S_k u_n)$ is also Cauchy for every $k \in \mathbf{N}$ and as e_k^* is continuous on the finite-dimensional subspace $S_k(X)$,

$$\lim_{n\to\infty} e_k^*(S_k u_n) = \alpha_k \quad \text{exists,} \quad k \in \mathbf{N}.$$

Similarly

$$\lim_{n\to\infty} S_k u_n = \sum_{i=1}^{k} \alpha_i e_i \qquad k \in \mathbf{N}$$

and this convergence is uniform in k for $\|\cdot\|$.

Hence $\sum_{i=1}^{\infty} \alpha_i e_i = u$ and so

$$\lim_{n\to\infty} S_k u_n = S_k u \quad \text{uniformly in } k.$$

But this means $|||u-u_n||| \to 0$.

Now the Closed Graph Theorem applies to the identity map $(X,\|\cdot\|) \to (X,|||\cdot|||)$ (or the Open Mapping Theorem to its inverse.) Hence $|||\cdot|||$ is continuous on X, and this implies the equicontinuity of (S_n) and the continuity of each e_n^*. □

An F-space with a basis must be separable. It is easy to use Theorem 1.8 to demonstrate that a separable F-space need not have a basis. Any F-space with no non-zero linear functionals (e.g. L_p, $0 < p < 1$, - see Chapter 2) cannot have a basis. Of course, the classical problem of the existence of a basis in any separable Banach space was solved negatively by Enflo [1972].

The proof of Theorem 1.8 can be extended for Banach spaces to give the Weak Basis Theorem of Mazur. Any basis of a Banach space (or locally convex F-space) in its weak topology is automatically a basis for the original topology. Recently, results of Shapiro [1974], Drewnowski [1977] and Morrow [1980] combined to show that the converse is true. Precisely, an F-space with a basis, in which every basis for the weak topology is a basis for the original topology, is necessarily locally convex.

CHAPTER 2
SOME OF THE CLASSICAL RESULTS

1. Introduction

In this chapter we introduce some of the spaces we will be working with, and we give some of their basic properties.

Most of the discussion in this chapter is about dual space questions. We check that each of the spaces we introduce is not locally convex. Then we try to see how large the dual space is: are there non-zero elements in the dual space? Are there enough continuous linear functionals to separate points? Is it possible to characterize the dual space?

For the Orlicz function spaces we introduce, we give a necessary and sufficient condition for the dual space to be non-trivial. We show that although ℓ_p, $0 < p < 1$, has a separating dual, it has a quotient space with trivial dual. The Banach envelope of X (which we introduce in connection with ℓ_p) is in a natural way a Banach space containing X and having the same dual space as X.

There is also some discussion of "how locally convex" a locally bounded space is. We show that although ℓ_p is p-convex, it is, in a strong sense, not q-convex for any $q > p$. Finally, we give the definition of a locally bounded space which is p-convex for all $p < 1$ but which is not locally convex.

The history of these particular examples goes back at least as far as 1931, when Nikodym proved that L_0 has trivial dual. Day's result that L_p, $0 < p < 1$, has trivial dual appeared in 1940. The notion of Banach envelope crystallized in the 1969 paper of Duren-Romberg-Shields. In the early 1970's Stiles studied the basic structure of ℓ_p, $0 < p < 1$.

We now turn to the examples.

2. The L_p spaces

Our first examples come from measure theory. Let (Ω,Σ,μ) be a finite measure space, and define $L_0(\mu)$ to be the space of all Σ-measurable functions on Ω, with the usual convention about identifying functions equal almost everywhere.

For $0 < p < \infty$, define $L_p(\mu)$ to be the space of functions f in $L_0(\mu)$ such that $\int |f(x)|^p d\mu(x) < \infty$. Define $L_\infty(\mu)$ to be the space of essentially bounded functions in $L_0(\mu)$, with the essential supremum norm

$$\|f\|_\infty = \text{ess sup } |f|.$$

The spaces $L_\infty(\mu)$ and $L_p(\mu)$, $p \geq 1$ are well-known Banach spaces.

The inequalities

$$(2.1) \qquad a+b/1+a+b \leq a/1+a + b/1+b, \quad a,b \geq 0$$

$$(2.2) \qquad (a+b)^p \leq a^p + b^p, \quad a,b \geq 0, \quad p \leq 1,$$

show that

$$\|f\|_0 = \int (|f(x)|/1+|f(x)|)\, d\mu(x)$$

and

$$\|f\|_p^p = \int |f(x)|^p d\mu(x)$$

define F-norms on $L_0(\mu)$ and $L_p(\mu)$, $0 < p \leq 1$ respectively. Note that the topology induced by the L_0 metric is just the topology of convergence in measure.

The companion inequality to 2.2 is

$$(2.3) \qquad (a+b)^p \leq 2^{p-1}(a^p+b^p) \quad \text{for } 1 \leq p.$$

For $0 < p \leq 1$, the function

$$\|f\|_p = \left(\int |f(x)|^p d\mu(x)\right)^{1/p}$$

is a quasi-norm on L_p. The function $\|f\|_p$ trivially satisfies equations 1.12 of Chapter 1; inequality 1.15 follows from 2.3:

$$\begin{aligned} \|f+g\|_p &= (\|f+g\|_p^p)^{1/p} \\ &\leq (\|f\|_p^p + \|g\|_p^p)^{1/p} \\ &\leq 2^{1/p-1}(\|f\|_p + \|g\|_p). \end{aligned}$$

We leave it to the reader to show that $L_0(\mu)$ is not locally bounded. The spaces $L_0(\mu)$ and $L_p(\mu)$ are complete and hence are F-spaces (see Kothe [1969]).

<u>Convention</u>. Whenever we use L_0 and L_p instead of $L_0(\mu)$, $L_p(\mu)$, we are referring to L_0 (respectively, L_p) of the unit interval with Lebesgue measure.

The <u>dual space</u> of a topological vector space X is the space of all continuous linear functionals on X; we denote this space by X^*. We say that X has <u>trivial dual</u> or is a <u>trivial dual space</u> if X^* consists only of the zero functional.

PROPOSITION 2.1. Let X be a Hausdorff topological vector space. Then

(i) $x^*(x) = 0$ for every $x^* \in X^*$ if and only if $x \in \text{co}\, U$ for every 0-neighborhood U;

(ii) X has trivial dual if and only if $X \subset \text{co}\, U$ for every 0-neighborhood U.

<u>Proof</u>. (ii) is an obvious consequence of (i). For (i): if $x \notin \text{co}\, U$ for some 0-neighborhood U, then by a standard separation theorem (see Kothe [1969], p. 192) there is a linear functional x^* on X which is 1 at x and bounded by 1 on U. Since x^* is bounded on U, it is continuous. Conversely, if $x^*(x) \neq 0$, the set

$$\{z \in X : |x^*(z)| < |x^*(x)|\}$$

is a convex 0-neighborhood not containing x. □

Now suppose A is an atom of (Ω,Σ,μ). Then $f \to f(A)$ (the a.e. constant value of f on A) is a continuous linear functional on $L_0(\mu)$ and on $L_p(\mu)$, $p > 0$; this is easy to check.

The next theorem takes care of the non-atomic case.

THEOREM 2.2. Assume μ is non-atomic.

(i) There is no non-zero operator from $L_0(\mu)$ into any locally bounded space;

(ii) There is no non-zero operator from $L_p(\mu)$, $0 < p < 1$, into a q-convex space, $q > p$.

In particular, L_0 and L_p, $0 < p < 1$, have trivial dual.

<u>Proof</u>. Without loss of generality we can assume $\mu(\Omega) = 1$. For the proof of (i), suppose T is an operator from L_0 into $(X,\|\cdot\|)$; by Theorem 1.3 we may assume $\|\cdot\|$ is r-convex for some r in $(0,1]$. Let $\epsilon > 0$ and let $U = \epsilon B_X$ and $V = T^{-1}(U)$. Note that V is r-convex.

By a standard exhaustion argument, for each $n \in \mathbf{N}$ there is a partition of Ω by sets E_i in Σ, $1 \leq i \leq n$, with $\mu(E_i) = 1/n$ for each i (recall μ is atomless). Now for any f in $L_0(\mu)$,

$$f = n^{-1/r} \sum_{i=1}^{n} n^{1/r} f 1_{E_i}.$$

For large enough n each function $n^{1/r} f 1_{E_i}$ is in V, since the supports of these functions are uniformly small. Thus f is written as an r-convex combination of elements of V, so $\|Tf\| \leq \epsilon$. Since ϵ and f are arbitrary, $T \equiv 0$.

The proof of (ii) requires only minor modifications. Assume T maps $L_p(\mu)$ into a q-convex space, $q > p$, and assume $f \in L_p(\mu)$ is <u>bounded</u>. Write

$$f = \sum_{i=1}^{n} f 1_{E_i},$$

where the E_i's are as before.

Then

$$\begin{aligned} \|Tf\|^q &\leq \sum_{i=1}^{n} \|T(f 1_{E_i})\|^q \\ &\leq \|T\|^q \sum_{i=1}^{n} \|f 1_{E_i}\|^q \\ &\leq \|T\|^q M^q n^{1-q/p} \\ &\qquad \text{(where } M \text{ is a bound for } f\text{)} \\ &\to 0 \quad \text{as} \quad n \to \infty. \end{aligned}$$

Thus T vanishes on the dense subspace of bounded functions, so T vanishes identically. □

See Theorem 2.12 for a similar result in a more general setting.

REMARKS. (1) Careful examination of the proof of Theorem 2.2 (i) shows the following: for any 0-neighborhood V in $L_0(\mu)$, there is an integer n so that $V + V + \ldots + V$ (n times) $=$ $L_0(\mu)$. From this it follows that <u>no non-zero operator from</u> $L_0(\mu)$

<u>into an</u> F-<u>space is compact</u>, since the equation immediately above shows that a compact operator has to map $L_0(\mu)$ into a compact set.

(2) The fact that $L_0 = L_0([0,1])$ has trivial dual is due to Nikodym [1931]. Day [1940] proved that for $0 < p < 1$ the dual of $L_p(\mu)$ is isometric to $\ell_\infty(A)$, where A is the collection of atoms of μ.

3. <u>The ℓ_p spaces</u>

For $0 < p < \infty$, define ℓ_p to be the space of all sequences $a = (a_n)$ such that

$$\|a\|_p^p = \Sigma|a_n|^p < \infty.$$

Define ℓ_∞ to be the space of sequences $a = (a_n)$ such that

$$\|a\|_\infty = \sup_n|a_n| < \infty.$$

The spaces ℓ_p, $p \geq 1$, and ℓ_∞ are well-known Banach spaces. From inequality 2.2, $\|\ \|_p^p$ is an F-norm on ℓ_p, $p < 1$, and from inequality 2.3, $\|\cdot\|_p$ is a quasi-norm, with modulus $\leq 2^{1/p-1}$. (It is easy to see that the modulus is precisely $2^{1/p-1}$.) It is a standard argument that ℓ_p is complete. An obvious but very important fact is that the sequence (e_i) is a basis for ℓ_p, where $e_i = (\delta_{in})$.

For $i \in N$, $|a_i| \leq \|a\|_p$, so $a \to a_i$ is a continuous linear functional on ℓ_p. Thus $\ell_p{}^*$ separates the points of ℓ_p. More is true: the dual of ℓ_p is isometric in a natural way to ℓ_∞.

To fix notation, let $(X, \|\cdot\|)$ be a locally bounded space, $\|\cdot\|$ a quasi-norm on X. Let B_X = unit ball of $X = \{x \in X : \|x\| \leq 1\}$. We can define a norm on X^* by

$$\|x^*\| = \sup\{|x^*(x)| : x \in B_X\}. \tag{2.4}$$

THEOREM 2.3. (Day [1940]) For $0 < p \leq 1$, ℓ_p^* is isometric to ℓ_∞.

<u>Proof</u>. For $x = (x_n)$ in ℓ_∞, define λ_x on ℓ_p by

$$\lambda_x(a) = \sum_{n=1}^{\infty} a_n x_n.$$

The above series converges and λ_x is continuous, since

$$|\Sigma a_n x_n| \leq \|x\|_\infty \Sigma |a_n|$$

$$\leq \|x\|_\infty \|a\|_p.$$

Thus λ_x is continuous and $\|\lambda_x\| \leq \|x\|_\infty$. Also, $\lambda_x(e_i) = x_i$, so $\|\lambda_x\| \geq \sup|x_i| = \|x\|_\infty$. Next, if λ is any element of ℓ_p^*, define $x = (x_i)$ by

$$x_i = \lambda(e_i).$$

The element x is in ℓ_∞ since $\sup|x_i| \leq \|\lambda\|$. Finally, $\lambda_x = \lambda$, since both are continuous linear functionals which agree on the finitely non-zero elements of ℓ_p. This shows that $x \to \lambda_x$ is an isometry of ℓ_∞ onto ℓ_p^*. □

Thus ℓ_p has many continuous linear functionals. However,

PROPOSITION 2.4. For $0 < p < 1$, ℓ_p is not locally convex.

Proof. For each i, $\|e_i\|_p = 1$. However,

$\|(1/n) \sum_{i=1}^{n} e_i\|_p = n^{1/p-1} \to \infty$ as $n \to \infty$, so the convex hull of B_{ℓ_p} is unbounded. Since ℓ_p is locally bounded, it follows that it has no bounded convex neighborhood of 0. □

The next group of results on the structure of ℓ_p is due to Stiles. Theorems 2.5 and 2.6 were proved for the case $p \geq 1$ by Pelczynski [1960].

THEOREM 2.5. (Stiles [1970]). For $0 < p < 1$, every closed infinite-dimensional subspace of ℓ_p contains a subspace isomorphic to ℓ_p.

Proof. This is a "gliding hump" argument. The two easy facts we require are

(i) Given x in ℓ_p and $\eta > 0$, there is $n \in N$ so that
$$\sum_{i=n+1}^{\infty} |x_i|^p < \eta;$$

(ii) If X is an infinite-dimensional subspace of ℓ_p, then for any m in N there is an x in X, $x \neq 0$, such that $x_i = 0$ for $1 \leq i \leq m$.

Now suppose X is an infinite-dimensional subspace of ℓ_p and $\epsilon > 0$. Using (i) and (ii) we can construct inductively a sequence (x^n) in X and a strictly increasing sequence of integers (i_n) such that

(iii) $\|x^n\| = 1$ for all n;

(iv) $x_j^n = 0$ for $j < i_{n-1}$;

(v) $\sum_{j=i_n}^{\infty} |x_j^n|^p < \epsilon$.

(Condition (iv) is vacuous for $n = 1$.)

We show that the closed span of (x^n) is isomorphic to ℓ_p. Let $\alpha_1,\ldots,\alpha_k$ be scalars. Then by p-convexity

(vi) $$\|\sum_{j=1}^{k} \alpha_j x^j\|^p \leq \sum_{j=1}^{k} |\alpha_j|^p.$$

On the other hand,

(vii) $$\|\sum_{j=1}^{k} \alpha_j x_j\|^p \geq \sum_{j=1}^{k} \sum_{i=n_{j-1}+1}^{n_j} |\sum_{j=1}^{k} \alpha_j x_i^j|^p$$

$$= |\alpha_1|^p \sum_{i=1}^{n_1} |x_i^1|^p + \sum_{i=n_1+1}^{n_2} |\alpha_1 x_i^1 + \alpha_2 x_i^2|^p$$

$$+ \sum_{i=n_2+1}^{n_3} |\alpha_1 x_i^1 + \alpha_2 x_i^2 + \alpha_3 x_i^3|^p \ldots$$

$$\geq |\alpha_1|^p(1-\epsilon) + |\alpha_2|^p(1-\epsilon) - |\alpha_1|^p \sum_{i=n_1+1}^{n_2} |x_i^1|^p$$

$$+ |\alpha_3|^p(1-\epsilon) - |\alpha_2|^p \sum_{i=n_2+1}^{n_3} |x_i^2|^p$$

$$- |\alpha_1|^p \sum_{i=n_2+1}^{n_3} |x_i^1|^p \ldots$$

$$\geq \left(\sum_{j=1}^{k} |\alpha_j|^p\right)(1-2\epsilon).$$

Now define T on the finitely non-zero elements of ℓ_p by setting $T(e_n) = x^n$ and extending linearly. Inequalities (vi) and (vii) say that if z is a finitely non-zero element of ℓ_p,

$$(1-2\epsilon)\|z\|_p^p \leq \|T(z)\|_p^p \leq \|z\|_p^p.$$

Thus T is an isomorphism on a dense subspace of ℓ_p, and so extends to be an isomorphism of ℓ_p into Y. Note too that $\|T\|\cdot\|T\|^{-1} \leq (1-2\epsilon)^{1/p}$, so we can choose T to be arbitrarily close to being an isometry. □

Stiles [1972] showed that in the proof of Theorem 2.5, if X is assumed to be complemented in ℓ_p, then the subspace of X constructed in the proof can be taken to be complemented in ℓ_p. Using this and the "Pelczynski decomposition method", he proved the following:

THEOREM 2.6. For $0 < p < 1$, every complemented subspace of ℓ_p is isomorphic to ℓ_p.

LEMMA 2.7. The quasi-norm on ℓ_p is not equivalent to a q-convex quasi-norm for any $q > p$.

<u>Proof</u>. Suppose the contrary. Then there would exist a constant C so that

$$n^{q/p} = \left\|\sum_{i=1}^{n} e_i\right\|^q \leq C\sum_{i=1}^{n} \|e_i\|^q = Cn \quad \text{for all } n,$$

an impossibility. □

From Theorem 2.5 and Lemma 2.7, we immediately deduce

COROLLARY 2.8. No infinite-dimensional subspace of ℓ_p is q-convex for any $q > p$.

PROPOSITION 2.9. (Stiles [1972]). If X is a q-Banach space every operator from X into ℓ_p is compact, for $p < q$.

Proof. We simply sketch the ideas. Suppose X is q-Banach and $T : X \to \ell_p$ is not compact. Then if U is the unit ball of X there is a sequence (u_n) in U such that the elements $v_n = T(u_n)$ are equivalent to the ℓ_p-basis. (This follows by an inductive argument using (i) and (ii) of the proof of Theorem 2.5.) But then the q-convex hull of the sequence v_n would have to be bounded, since U is q-convex, and this contradicts Lemma 2.7. □

In a locally convex t.v.s., every closed subspace is closed in the weak topology, by the separation theorem. The space ℓ_p, $0 < p < 1$, is the simplest example of a locally bounded space whose dual separates points and yet which has a closed subspace dense in the weak topology. This follows from the fact that ℓ_p is a "universal covering space" for separable p-Banach spaces:

PROPOSITION 2.10. Let $(X, \|\ \|)$ be a separable p-Banach space. Then there is a continuous linear map from ℓ_p onto X.

Proof. Let (x_i) be a sequence dense in B_X, and let (e_i) be the usual unit basis vectors in ℓ_p. Define $T : \ell_p \to X$ by

$$T\left(\sum_{i=1}^{\infty} \alpha_i e_i\right) = \sum_{i=1}^{\infty} \alpha_i x_i.$$

The series defining T converges, since if $j,k \in N$, $j < k$, then

$$\|\sum_{i=j}^{k} \alpha_i x_i\|^p \le \sum_{i=j}^{k} \|\alpha_i x_i\|^p \quad \text{(by p-convexity)}$$

$$\le \sum_{i=j}^{k} |\alpha_i|^p.$$

The same computation shows that T is continuous.

Since (x_i) is dense in B_X, $T(B_{\ell_p})$ is dense in B_X; Theorem 1.4 implies that T is onto.

Using the preceding result, Shapiro [1969] and Stiles [1970] constructed a proper closed weakly dense subspace of ℓ_p.

COROLLARY 2.11. For $0 < p < 1$, ℓ_p has a proper closed weakly dense subspace.

<u>Proof</u>. The space L_p is a separable p-Banach space, so by Proposition 2.10 there is a continuous linear surjection $T : \ell_p \longrightarrow L_p$. Let $M = \ker T$; then by the open mapping theorem ℓ_p/M is isomorphic to L_p. The subspace M is closed, and $(\ell_p/M)^* = L_p^* = \{0\}$. So any continuous linear functional on ℓ_p which vanishes on M is identically zero, since otherwise it induces a non-zero element of $(\ell_p/M)^*$. Thus M is weakly dense. □

4. The Banach envelope

As simple geometric motivation, consider the plane with the ℓ_p norm, $0 < p < 1$, and with the ℓ_1 norm. The identity map from $(R^2, \|\cdot\|_p)$ to $(R^2, \|\cdot\|_1)$ is continuous. The two spaces are equal as sets (in this case). The convex hull of the ℓ_p unit ball is the ℓ_1 unit ball. The space $(R^2, \|\cdot\|_1)$, then, is the "smallest" Banach space containing $(R^2, \|\cdot\|_p)$.

DEFINITION. Let $(X, \|\cdot\|)$ be a locally bounded space whose dual separates points. The containing Banach space of X, or Banach envelope of X, $\hat{X}$, is the completion of the quasi-normed space $(X, \|\cdot\|_c)$, where $\|\cdot\|_c$ is the gauge functional of $\operatorname{co} B_X$.

Note that $\|\cdot\|_c$ is actually a norm on X. To see this, suppose $x \in X$ and $x \neq 0$. Then by condition (i) of Proposition 2.1 there exists $\epsilon > 0$ such that $x \notin \operatorname{co}(\epsilon B_X) = \epsilon \operatorname{co} B_X$, i.e., $\|x\|_c \geq \epsilon$.

Thus the identity map $i : (X, \|\ \|) \to (X, \|\ \|_c)$ is continuous (it is norm one), and by construction its range is dense in $\hat{X}$.

Finally, note that X and $\hat{X}$ have the same dual space and the dual norms are the same. For if x^* is in X^*, then $\|x^*\| = \sup_{x \in B_X} |x^*(x)| = \sup_{z \in \operatorname{co} B_X} |x^*(z)|$. Thus x^* is $\|\cdot\|_c$-continuous on X, and may be extended to $\hat{X}$ with the same norm. Conversely, every element of $\hat{X}^*$ is in X^*.

Another way of constructing $\hat{X}^*$ is as follows: the space X^* is a Banach space with the norm defined in (2.4). Define $J : X \to X^{**}$ by $(Jx)(x^*) = x^*(x)$. Note that the norm of J is at most 1, and J is 1-1 since X^* separates the points of X. Then the Banach envelope of X^* can be taken to be the closure in X^{**} of the image of X under J.

This construction is equivalent to the one in the definition, since for any $x^* \in X^*$,

$$\|x^*\| = \sup_{x \in B_X} |x^*(x)| = \sup_{x \in \mathrm{co}\, B_X} |x^*(x)|.$$

From this

$$\|Jx\| = \sup_{\|x^*\| \leq 1} |x^*(x)| = \|x\|_c .$$

There is an "isomorphic" version of the construction of $\hat{X}$. Let Y be a Banach space containing the locally bounded space X as a dense subspace. Suppose $B_X \subset B_Y$. Suppose further that for some $C > 0$, $X \cap B_Y \subset C\, \overline{\mathrm{co}}\, B_X$. Then Y is isomorphic to X (with isomorphism constant $\leq C$).

To illustrate the ideas, we verify that the containing Banach space of ℓ_p, $0 < p < 1$, is ℓ_1. First note that ℓ_p is a dense subspace of ℓ_1, and that $B_{\ell_1} \subset B_{\ell_p}$. Now if x is a finitely non-zero element of ℓ_p with $\|x\|_1 \leq 1$, then $x \in \mathrm{co}\, B_{\ell_p}$. Let F be the space of finitely non-zero elements of ℓ_1; then $(\mathrm{co}\, B_{\ell_p}) \cap F = B_{\ell_1} \cap F$. Taking the completion, we see that ℓ_1 is the envelope of ℓ_p.

5. Orlicz function spaces and sequence spaces

Other function spaces and sequence spaces whose topology is given by a Δ-norm are provided by the construction we now outline. These spaces - Orlicz function and sequence spaces - occur frequently in the literature, particularly as useful examples. We will not do a careful study of these spaces in these notes, but it should be pointed out that many of our results carry over to this more general setting. For instance, with only minor wording changes, the construction in Chapter 9 can be carried out in suitable Orlicz function spaces.

Let φ be a real-valued function defined on $[0,\infty)$, satisfying

(2.5) $\qquad \varphi(t) = 0$ if and only if $t = 0$, and
φ is continuous at 0

(2.6) $\qquad \varphi$ is increasing

(2.7) $\qquad \varphi(2t) \leq C\varphi(t)$ for all t and some constant C.

A function φ as above is an <u>Orlicz function</u>. (Note: many authors define an Orlicz function as a function satisfying only 2.5 and 2.6. Condition 2.7 is known as the Δ_2-condition, and is then added as an extra condition.)

Note that for all $0 \leq s,t$,

$$(2.8) \qquad \begin{aligned} \varphi(s+t) &\leq \varphi(2\max(s,t)) \\ &\leq C\varphi(\max(s,t)) \\ &\leq C(\varphi(s)+\varphi(t)). \end{aligned}$$

Now let (Ω,Σ,μ) be a finite measure space and φ an Orlicz function. Define

$$L_\varphi(\mu) = \{f \in L_0(\mu) : \int \varphi(|f(x)|)d\mu(x) < \infty\}.$$

Conditions 2.5, 2.6, and 2.8 show that

$$\|f\|_\varphi = \int \varphi(|f(x)|)d\mu(x)$$

satisfies (1.1)-(1.4) in the definition of Δ-norm. Condition (1.3) follows immediately from the continuity of φ at 0 and the dominated convergence theorem. The space L_φ is the <u>Orlicz space</u> determined by φ.

The next theorem is due to Rolewicz [1959]; see also Cater [1963].

THEOREM 2.12. Let (Ω,Σ,μ) be a finite measure space, and let φ be an Orlicz function. Assume that μ is non-atomic. Then $L_\varphi(\mu)$ has trivial dual if and only if $\underline{\lim}_{t\to\infty} (\varphi(t))/t = 0$.

Proof. The "if" part follows essentially as in Theorem 2.2. For the "only if" part, assume $\underline{\lim}_{t\to\infty} (\varphi(t)/t) \neq 0$. We will show that L_φ embeds continuously in L_1, whence $L_\varphi{}^*$ will, in fact, separate the points of L_φ.

Choose $\beta > 0$ with $(\varphi(t)/t) \geq \beta$ for $t \geq 1$. Now suppose (f_n) is a sequence in $L_\varphi(\mu)$ with $\lim_{n\to\infty} \|f_n\|_\varphi = 0$. Let $E_n = \{x \in \Omega : |f_n(x)| \leq 1\}$. Since $\|f_n\|_\varphi \to 0$, $\varphi(|f_n|)$ converges to 0 in measure so $|f_n|$ converges to 0 in measure, and then $\lim_{n\to\infty} \int |f_n| 1_{E_n} \, d\mu = 0$. Also

$$|f_n 1_{\sim E_n}| \leq 1/\beta \, \varphi(f_n) 1_{\sim E_n},$$

so $\lim_{n\to\infty} \int |f_n| 1_{\sim E_n} \, d\mu = 0$. These calculations show that $L_\varphi(\mu)$ is a subset of $L_1(\mu)$ and that the identity map from $L_\varphi(\mu)$ to $L_1(\mu)$ is continuous. □

We state without proof two further results on the structure of Orlicz function spaces:

THEOREM 2.13. (Rolewicz [1959]). Let φ be an Orlicz function. A necessary and sufficient condition that L_φ be locally bounded is that

$$\lim_{\lambda\to 0} \lim_{t\to\infty} (\varphi(\lambda t)/\varphi(t)) = 0.$$

THEOREM 2.14. (Matuszewska and Orlicz [1961]). Let φ be an Orlicz function and let $p \in (0,1]$. Then L_φ is p-convex if and only if there is a p-convex Orlicz function γ so that $L_\varphi = L_\gamma$.

Let φ be an Orlicz function. The discrete analogue of L_φ is

$$\ell_\varphi = \{(a_n) : \sum_n \varphi(|a_n|) < \infty\},$$

the Orlicz sequence space defined by φ. The dual space of ℓ_φ separates the points of ℓ_φ, since $a \to a_n$ is a continuous linear functional for each n. (This follows easily from the fact that φ is increasing and $\varphi(t) = 0 \implies t = 0$.)

It is easy to check that if $\|\cdot\|$ is a p-convex quasi-norm on X, it is q-convex for all $q < p$. The usual quasi-norm on ℓ_p is p-convex; we remarked before 2.6 that it is not equivalent to a q-convex quasi-norm for any $q > p$.

A sequence space whose quasi-norm shows a different type of behavior was considered by Bourgin [1943]. For each n, let

$$p_n = (1 + (\log(n+1))^{-1/2})^{-1},$$

and define

$$\ell_{(p_n)} = \{(a_n) : \sum_{n=1}^{\infty} |a_n|^{p_n} < \infty\}.$$

The sum $\sum|a_n|^{p_n}$ defines an F-norm on $\ell_{(p_n)}$. Bourgin shows that for every $p < 1$, $\ell_{(p_n)}$ can be given an equivalent p-convex quasi-norm; yet $\ell_{(p_n)}$ is not locally convex.

Let (p_n) be <u>any</u> sequence increasing to 1, and define $\ell_{(p_n)}$ as above. A necessary and sufficient condition that $\ell_{(p_n)}$ be locally convex is that

$$\sum_{n=1}^{\infty} N^{p_n/(p_n-1)} < \infty \text{ for some integer } N > 1;$$

see Simons [1965].

CHAPTER 3
HARDY SPACES

1. Introduction

Much of the early impetus to develop an adequate theory of non-locally convex spaces has come from work in the classical Hardy spaces, H^p. Because there is no clear intrinsic reason to restrict attention to the case $p \geq 1$, the theory of the spaces H^p for $p < 1$ has been studied in some detail, and so these provide some of the most interesting and best understood examples of non-locally convex spaces.

We shall denote by D the open unit disc in the complex plane and by T the unit circle. For $0 < p \leq \infty$ we denote by H^p the space of analytic functions $f : D \to C$ so that for $0 < p < \infty$)

$$(3.1) \qquad \|f\|_p = \sup_{0 \leq r < 1} (1/2\pi \int_0^{2\pi} |f(re^{i\theta})|^p d\theta)^{1/p} < \infty$$

or, for $p = \infty$,

$$\|f\|_\infty = \sup_{z \in D} |f(z)|.$$

It is clear that if $1 \leq p \leq \infty$, $\|\cdot\|_p$ is a norm on H^p, while if $0 < p < 1$, $\|\cdot\|_p^p$ is an F-norm, so that $\|\cdot\|_p$ is a p-subadditive quasi-norm.

The basic theory of the H^p spaces can be found in Duren [1970]. We shall simply quote the relevant information. If $f \in H^p$, for $p > 0$, then for almost every $\theta \in [-\pi,\pi]$

$$\lim_{r\to 1} f(re^{i\theta}) = f^*(e^{i\theta}) \tag{3.2}$$

exists. Furthermore

$$\lim_{r\to 1} \int_0^{2\pi} |f^*(e^{i\theta})-f(re^{i\theta})|^p d\theta = 0. \tag{3.3}$$

Let L_p denote the L_p-space of $\mathbf{T}$ with normalized Haar measure on the circle, i.e. $(2\pi)^{-1}d\theta$. Then if $f \in H^p$ and $r < 1$, $f_r \in L_p(\mathbf{T})$ where

$$f_r(e^{i\theta}) = f(re^{i\theta}).$$

Now $\|f_r\|_p$ is an increasing function of r and by (3.2) and (3.3) $\|f_r-f^*\|_p \to 0$. Hence the map $f \to f^*$ is an isometry of H^p into $L_p(\mathbf{T})$. It will be convenient to identify H^p as a closed subspace of $L_p(\mathbf{T})$ this way. Hence if $f \in L_p(\mathbf{T})$ we shall speak of $f \in H^p$ meaning that f is the boundary value of some function, also denoted by f, in H^p.

At this point it is not clear that H^p is either complete or that it is not isomorphic to L_p. These are both however established once we have the inequality

$$(3.4) \qquad |f(re^{i\theta})| \leq 2^{1/p}\|f\|_p(1-r)^{-1/p}.$$

From (3.4) we see that point evaluations $f \to f(z)$ for $z \in D$ are continuous, so that H^p has a separating dual. Completeness of H^p also follows from (3.4) since if (f_n) is a Cauchy sequence then it is a normal family on D. Thus for $0 < p < 1$, H^p is a p-Banach space, while for $1 \leq p \leq \infty$, H^p is a Banach space.

We shall also need to observe that the polynomials form a dense subspace of H^p for $0 < p < \infty$. In fact for each $r < 1$, $f_r \in H^p$ and $\|f_r - f\|_p \to 0$. However f_r is the uniform limit of polynomials for each $r < 1$.

A most important technique in the theory of H^p-spaces is the idea of <u>inner-outer factorization</u>. We need some notation. A function $S \in H_\infty$ is called <u>inner</u> if

$$|S(z)| \leq 1 \qquad z \in D$$

$$|S(e^{i\theta})| = 1 \qquad \text{a.e.}$$

If $F \in H^p$ then F is <u>outer</u> if

$$(3.5) \qquad F(z) = e^{i\gamma}\exp\left(1/2\pi \int_0^{2\pi} \frac{e^{it}+z}{e^{it}-z} \log\psi(t)\,dt\right)$$

where $\log\psi \in L_1$, $\psi \in L_p$ and $\gamma \in \mathbb{R}$.

Then the Canonical Factorization Theorem (Duren, p. 24) states that every $f \in H^p$ can be factored in the form $f = FS$

where F is an outer function and S is an inner function. Clearly $\|f\|_p = \|F\|_p$.

The inner function S may be further factorized. If (a_n) is a sequence in D (possibly finite) so that

$$\Sigma(1-|a_n|) < \infty$$

then the Blaschke product B associated to (a_n) is the function

$$(3.6) \qquad B(z) = \prod_{n=1}^{\infty} \frac{|a_n|}{a_n} \frac{a_n - z}{1-\bar{a}_n z} \qquad z \in D$$

(where $|a_n|/a_n$ is interpreted to be -1 if $a_n = 0$). B is also an inner function and has zeros at each a_n (cf. Duren, p. 19). If S is an inner function without zeros then S is called a singular inner function and takes the form

$$(3.7) \qquad S(z) = \exp(-\int_0^{\pi} (e^{it}+z/e^{it}-z)\, d\mu(t))$$

where μ is a positive singular measure (with respect to Lebesgue measure) on $(-\pi,\pi]$.

Now every inner function S can be factored in the form $S = BS_1$ where B is a Blaschke product and S_1 is a singular inner function.

THEOREM 3.1. If $f \in H^p$ is non-zero then f has a unique factorization $f = BSF$ where B is a Blaschke product, S is a singular inner function and F is an outer function in H^p.

REMARKS. Of course $B \equiv 1$ or $S \equiv 1$ are possible. Note that in (3.5) we must take $\psi(t) = |f(e^{it})|$; it is important to note that if $f \in H^p$ and $f \neq 0$ then

$$\int_{-\pi}^{\pi} \log|f(e^{it})|dt > -\infty.$$

2. Linear topological properties of H^p, $0 < p < 1$

Our first observation is an old result due to Livingston (1953) and Landsberg (1956).

THEOREM 3.2. If $0 < p < 1$, H^p is not locally convex.

Proof. The trigonometric polynomials are dense in $L_p(\mathbf{T})$. Hence given $\epsilon > 0$ and $n \in \mathbf{N}$ we can find trigonometric polynomials $\varphi_1,\ldots,\varphi_n$ satisfying $\|\varphi_i\|_p \leqslant 1$ and

$$\|\varphi_1+\ldots+\varphi_n\|_p \geqslant (1-\epsilon)n^{1/p}.$$

Pick $N \in \mathbf{N}$ so large that $z^N\varphi_i \in H^p$ for $1 \leqslant i \leqslant n$. Then if $\psi_i = z^N\varphi_i$, $\|\psi_i\|_p \leqslant 1$ and

$$\|1/n(\psi_1+\ldots+\psi_n)\|_p \geqslant (1-\epsilon)n^{1/p-1}$$

i.e. H^p is not locally convex. □

We have already observed that H^p has a separating dual and is thus not isomorphic to L_p.

THEOREM 3.3 H^p contains a complemented subspace isomorphic to ℓ_p.

Proof. Let (a_k) be a uniformly separated sequence in D i.e. for some $\delta > 0$,

$$\prod_{\substack{j=1\\ j\neq k}}^{\infty} |a_k - a_j/1-\bar{a}_j a_k| \geq \delta$$

for every k. Then the map $T : H^p \to \ell_p$ defined by

$$(Tf)_n = (1-|a_n|^2)^{1/p} f(a_n)$$

is an open mapping of H^p onto ℓ_p. (This is a deep result - see Duren p. 149.) Thus we can find $g_n \in H^p$ with $\sup\|g_n\|_p < \infty$ and $Tg_n = e_n$. If $S : \ell_p \to H^p$ is defined by

$$S(t) = \sum_{n=1}^{\infty} t_n g_n$$

then S is bounded and ST is a projection of H^p onto a subspace isomorphic to ℓ_p. □

In view of this we next establish that H^p is not isomorphic to ℓ_p.

THEOREM 3.4. H^p contains a subspace isomorphic to ℓ_2.

<u>Proof</u>. In fact by a theorem of Paley (Duren, p. 104) the closed linear span M of the sequence $(z^{n_k} : k = 1,2,\ldots)$ is isomorphic to ℓ_2 when n_k is any lacunary sequence of integers i.e.

$$\inf_k n_{k+1}/n_k > 1.$$

Let us give a simple proof for the special case $n_k = 3^k$.

Suppose

$$f(z) = \sum_{k=1}^{m} a_k z^{n_k} .$$

Then

$$\|f\|_2^2 = \sum_{k=1}^{m} |a_k|^2$$

while

$$\|f\|_4^4 = 1/2\pi \int_{-\pi}^{\pi} \sum_{h,j,k,\ell \leq m} a_h a_j \bar{a}_k \bar{a}_\ell e^{i(n_h+n_j-n_k-n_\ell)\theta} dt$$

$$= 2 \sum_{j \neq k} |a_j|^2 |a_k|^2 + \sum_{k=1}^{m} |a_k|^4$$

$$\leq 2\|f\|_2^4$$

where the second equality depends on the fact that $n_h+n_j = n_k+n_\ell$ implies that $h = k$ or $h = \ell$.

Now if α is chosen so that $\alpha/p + (1-\alpha)/4 = 1/2$ then by Holder's inequality

$$\|f\|_2 \leq \|f\|_p^{\alpha} \|f\|_4^{1-\alpha}$$

$$\|f\|_2^{\alpha} \leq 2^{1/4} \|f\|_p^{\alpha}$$

i.e. $$\|f\|_2 \leq 2^{1/(4\alpha)} \|f\|_p .$$

Since $\|f\|_p \leq \|f\|_2$, the H^p-norm on M is equivalent to the H^2-norm, i.e., $M \approx \ell_2$. □

Now as ℓ_p contains no subspace isomorphic to ℓ_2, we observe that $\ell_p \not\approx H^p$.

3. <u>The Banach envelope of</u> H^p

In this section we identify the Banach envelope of H^p with a certain Bergman space B_p of analytic functions on D. This result is due to Duren, Romberg, and Shields [1969]; however our approach is due to Shapiro [1976].

We start with an inequality due to Hardy and Littlewood (Duren, p. 87).

THEOREM 3.5. There is a constant C depending only on p so that if $f \in H^p$

$$\int_0^1 (1-r)^{1/p-2} \|f_r\|_1 dr \leq C\|f\|_p. \tag{3.8}$$

The proof of Theorem 3.5 is quite cumbersome and we refer the reader to Duren. □

Motivated by (3.8) we define B_p to be the space of functions f analytic on D such that

$$\|f\|_{p,1} = 1/\pi \int_{-\pi}^{\pi} \int_0^1 |f(re^{i\theta})|(1-r)^{1/p-2} r dr d\theta < \infty.$$

B_p is a normed space and if $f \in H^p$ then

$$\|f\|_{p,1} \leq C\|f\|_p$$

for some constant C depending only on p.

To show that B_p is complete we observe that if

$\|f\|_{p,1} \leq 1$ then for any $s < 1$

$$\|f_s\|_1 \leq \gamma_s$$

where

$$\gamma_s^{-1} = \int_s^1 2r(1-r)^{1/p-2}dr.$$

Thus the unit ball of B_p is a normal family on D. Hence if (f_n) is a Cauchy sequence then (f_n) can be shown to converge in B_p-norm to any of its analytic cluster points (in the topology of uniform convergence on compacta). Thus B_p is actually a Banach space, and of course point evaluations on D are continuous.

If $f \in B_p$, denote by $f_r \in B_p$ the function

$$f_r(z) = f(rz) \qquad z \in D, \quad 0 < r < 1.$$

Then for $0 < s < 1$,

$$\|f-f_s\|_{p,1} = \int_0^1 \|f_r-f_{rs}\|_1 \, 2r(1-r)^{1/p-2}dr$$

$$\to 0$$

as $s \to 1$, by the Dominated Convergence Theorem, since $\|f_r-f_{rs}\|_1 \to 0$ for each r and $\|f_r-f_{rs}\|_1 \leq 2\|f_r\|_1$. As $f_s \in H_\infty \subset H^p$ for each s, we see that H^p is dense in B_p.

Thus B_p is a completion of H^p for the norm

$\|\cdot\|_{p,1}$. To show that B_p is a Banach envelope of H^p requires us to show that the unit ball of B_p is contained in a multiple of the convex hull of the unit ball of H^p; Theorem 3.5 will give us the information that the unit ball of B_p is a neighborhood of zero in H^p.

It will be convenient to introduce $\sigma = 1/p - 2$ and let β be an integer so that $\beta > \sigma$. Define for $z,w \in D$

$$(3.9) \qquad K(z,w) = (\beta+1)\,(1-|w|^2)^{\beta}\cdot(1-\bar{w}z)^{-\beta-2})$$

$$(3.10) \qquad J_w(z) = (1-|w|)^{-\sigma}K(z,w).$$

LEMMA 3.6. The set $(J_w : w \in D)$ is bounded in H^p.

Proof. For $0 < r < 1$

$$\int_0^{2\pi} |J_w(re^{i\theta})|^p \, d\theta/2\pi$$

$$= \int_0^{2\pi} \left(\frac{(1-|w|)^{\beta-\sigma}(1+|w|)^{\beta}}{|1-r\bar{w}e^{i\theta}|^{\beta+2}}\right)^p d\theta/2\pi$$

$$\leq 2^{\beta p}(1-|w|)^{p(\beta-\sigma)} \int_0^{2\pi} |1-r\bar{w}e^{i\theta}|^{-(\beta+2)p} \, d\theta/2\pi$$

$$\leq C(1-|w|)^{p(\beta-\sigma)}(1-|w|)^{1-(\beta+2)p}.$$

The last estimate can be deduced from the Lemma on p. 65 of Duren, since $(\beta+2)p > 1$. Thus

$$\|J_w\| \leq C^{1/p}(1-|w|)^{1/p-\sigma-2} = C^{1/p}. \quad \square$$

LEMMA 3.7. If φ is a polynomial,

$$\varphi(z) = 1/\pi \int_0^{2\pi} \int_0^1 \varphi(re^{i\theta})K(z,re^{i\theta})rdrd\theta.$$

Proof. It suffices to consider $\varphi(z) = z^n$.

$$1/\pi \int_0^1 \int_0^{2\pi} r^n e^{ni\theta} K(z,re^{i\theta})rdrd\theta$$

$$= (\beta+1)/\pi \int_0^1 (1-r^2)^{\beta} r^{n+1} I(r)dr$$

where

$$I(r) = \int_0^{2\pi} e^{in\theta}/((1-re^{-i\theta}z)^{\beta+2})\, d\theta$$

$$= \sum_{k=0}^{\infty} \int_0^{2\pi} ((\beta+2)\ldots(\beta+k+1))/k!\; r^k z^k e^{i(n-k)\theta} d\theta$$

$$= 2\pi\; (((\beta+n+1)!)/(\beta+1)!n!)\; r^n z^n.$$

Now

$$(\beta+1)/\pi \int_0^1 (1-r^2)^{\beta} r^{n+1} I(r)dr$$

$$= 2z^n\; ((\beta+n+1)!)/(\beta!n!) \int_0^1 (1-r^2)^{\beta} r^{2n+1} dr$$

$$= z^n.$$

The last equality is obtained by noting that

$$2\int_0^1 (1-r)^{2\beta} r^{2n+1} dr = \int_0^1 (1-u)^{\beta} u^n du$$

$$= ((\beta+n+1)!)/(\beta!n!) \qquad \text{(by induction)}. \quad \square$$

THEOREM 3.8. B_p is the Banach envelope of H^p.

Proof. Suppose $f \in H^p$ and $\|f\|_{p,1} < 1$. Then there is a polynomial φ so that

$$\|f-\varphi\|_p < (1/2)^{1/p},$$

$$\|\varphi\|_{p,1} < 1.$$

Now

$$\varphi(z) = \int_D J_w(z)\varphi(w)(1-|w|)^{\sigma} d\lambda(w)$$

where $d\lambda = r dr d\theta/\pi$. If $0 < s < 1$,

$$\varphi_s(z) = \int_D J_w(sz)\varphi(w)(1-|w|)^{\sigma} d\lambda(w).$$

Now

$$\int_D |\varphi(w)|(1-|w|)^{\sigma} d\lambda(w) = \|\varphi\|_{p,1}.$$

Since $(w,z) \to J_w(sz)$ is clearly continuous on $\bar{D} \times \bar{D}$, it follows that there is a convex combination

$$g(z) = \sum_{i=1}^{m} a_i J_{w_i}(sz)$$

where $\sum |a_i| = 1$ and

$$|\varphi_s(z)-g(z)| \leq (1/2)^{1/p} \qquad z \in D.$$

For large enough s,

$$\|f-\varphi_s\|_p < (1/2)^{1/p}$$

and so

$$\|f-g\|_p < 1.$$

Thus

$$f = (f-g) + \sum_{i=1}^{m} a_i J_{w_i}(sz)$$

$$= \sum_{i=0}^{m} c_i h_i$$

where $\|h_i\|_p \leq 1$ and $c_i \geq 0$ and

$$\sum_{i=0}^{m} c_i \leq 1 + C$$

where $C = \sup_i \|J_{w_i}\|_p$.

This, combined with Theorem 3.5, completes the proof since it shows that the unit ball of B_p is contained in $(C+1)\text{co}\, U$ where U is the unit ball of H^p. □

The kernel K defined in (3.9) is an example of a <u>reproducing kernel</u> and can be used to produce a projection of $L_1(\mu)$ onto B_p where μ is the measure on D given by

$$d\mu = (1-|w|)^{\sigma} d\lambda.$$

In fact if $f \in L_1(\mu)$ we define $Pf \in B_p$ by

$$Pf(z) = \int_D K(z,w)f(w)d\lambda(w)$$
$$= \int_D J_w(z)f(w)d\mu(w).$$

Then P is a projection of $L_1(\mu)$ onto its subpsace B_p.

Since the unit ball of B_p is a normal family of analytic functions on D, it is not difficult to see that B_p is isomorphic to a dual Banach space. Now a theorem due to Lewis and Stegall (1973) asserts that a complemented subspace of L_1 which is a dual Banach space is isomorphic to ℓ_1. Thus we have a theorem, originally due to Lindenstrauss and Pelczynski (1971):

THEOREM 3.9. B_p is isomorphic to ℓ_1.

Since B_p has an unconditional basis, it was a natural question to ask whether H^p has an (unconditional) basis. Note, of course, that the natural choice $\{z^n : n \geq 0\}$ is not a basis for H_p; consider the Taylor series expansion of $(1-z)^{-1}$. Very recently, Wojtasczyk [to appear] has answered this question by constructing an unconditional basis for H^p where $0 < p < 1$, using spline functions. It would take us too far afield to describe this construction here. The corresponding question for $p = 1$ was originally settled (again positively) by Maurey [1980] and Carleson [1980], Wojtasczyk [to appear].

It may be noted that both ℓ_p and H^p have ℓ_1, isomorphically, as their containing Banach space. We shall see later that the containing Banach space of a non-locally convex quasi-Banach space generally must exhibit a certain degree of ℓ_1-like behavior; precisely, ℓ_1 must be finitely representable in it.

PROBLEM 3.1. Classify the complemented subpsaces of H^p.

Every complemented subspace of H^p has a Banach envelope isomorphic to ℓ_1, since its containing Banach space can be identified with a complemented subspace of B_p. We have seen that ℓ_p can arise as such a complemented subspace. Kalton and Trautman [1982] have shown that no complemented subspace can be q-convex for any $q > p$, and this implies that every complemented copy of H^p contains a copy of ℓ_p. Wojtasczyk [to appear] shows that this copy can be chosen to be complemented.

4. PCWD subspaces of H^p

In Chapter 2, we saw that the space ℓ_p for $0 < p < 1$ has a proper closed subspace which is dense in the weak topology. This was demonstrated by showing that ℓ_p has a quotient space (L_p) which has trivial dual. Since ℓ_p is isomorphic to a complemented subspace of H^p it follows that H^p also has L_p as a quotient space and thus has a PCWD-subspace. However it is of some interest to show that H_p has PCWD-subspaces satisfying certain invariance properties. There are two types of invariance to consider. We say that a subspace X of H_p is <u>rotation-invariant</u> if whenever $f \in X$ then $f_w \in X$ for $|w| = 1$

where

$$f_w(z) = f(wz) \qquad |z| < 1.$$

We say X is <u>invariant</u> if $f \in X$ implies that $zf(z) \in X$.

We first consider rotation-invariant subspaces. For $p \geq 1$ the only rotation-invariant closed subspaces of H^p are those of ths form $H^p(M)$ where M is a subset of the non-negative integers. Here $f \in H^p(M)$ iff $\hat{f}(n) = 0$, $n \notin M$, where

$$\hat{f}(n) = 1/2\pi \int_0^{2\pi} f(e^{i\theta})e^{-in\theta}d\theta$$

$$= f^n(0)/n!.$$

Now $H^p(M)$ is also the closed linear span of the set $\{z^n : n \in M\}$.

De Leeuw [1940] first observed that for $p < 1$, there is a closed rotation-invariant subspace $J_p{}^0$ of H^p which fails spectral synthesis, i.e., contains none of the characters $(z^n : n \geq 0)$. To define this space consider H^p as a closed subspace of $L_p(T)$. We define $\bar{H}^p$ to be the set of $f \in L_p(T)$ such that $\bar{f} \in H^p$ where

$$\bar{f}(e^{i\theta}) = \overline{f(e^{i\theta})}.$$

$\bar{H}^p$ is also a closed linear subspace of L_p; in fact it is the closed linear span of $\{z^n : n \leq 0\}$. We shall define $J_p = H^p \cap \bar{H}^p$ and $J_p{}^0 = H^{p,0} \cap \bar{H}^p$ where

$H^{p,0} = \{f \in H^p : f(0) = 0\}$.

For $p > 1$, J_p consists only of the constants and $J_p^0 = \{0\}$. However for $p < 1$, H^p contains non-constant functions which are real on the boundary e.g.

$$f(z) = i\left(\frac{1+z}{1-z}\right),$$

and such functions belong to J_p. More generally it is not difficult to show that $z(1-z)^{-1} \in J_p^0$ and hence J_p^0 is clearly a rotation-invariant subspace containing no characters.

We will show that J_p is in fact weakly dense in H^p. We show this by proving a recent result of Aleksandrov ([1978],[1979]).

THEOREM 3.10. (Aleksandrov) $H^p + \overline{H}^p = L_p(\mathbb{T})$ for $0 < p < 1$.

<u>Remarks</u>. For $p > 1$ the same conclusion is true and follows from the fact that H^p is complemented in L_p. (See, for example, Duren.) For $p = 1$, the conclusion of the theorem is false.

<u>Proof</u>. Let us define the map $A : H^p \oplus \overline{H}^p \to L_p(\mathbb{T})$ by $A(f,g) = f+g$. It is clear that A has dense range; we need to show that A is a surjection. To do this it suffices to show that, if U is a 0-neighborhood in $H^p \oplus \overline{H}^p$, then $\overline{A(U)}$

contains a 0-neighborhood in L_p. Thus it will suffice to show that for some $C < \infty$, if $\varphi \in L_p(T)$ is a trigonometric polynomial with $\|\varphi\|_p = 1$ then there exist $f \in H^p$, $g \in \overline{H}^p$ with $\varphi = f+g$ and $\|f\|_p$, $\|g\|_p \leq C$.

Let us suppose $\|\varphi\|_p = 1$, where

$$\varphi(e^{i\theta}) = \sum_{k=-n}^{n} a_k e^{ik\theta} \qquad -\pi < \theta \leq \pi.$$

For $-\pi < \alpha \leq \pi$ define $f_\alpha \in H^p$ by

$$f_\alpha = -e^{i\alpha}/(1-e^{i\alpha}z^n) \sum_{k=0}^{2n} a_{k-n} z^k.$$

Then

$$\|f_\alpha\|_p^p = (1/2\pi) \int_0^{2\pi} \frac{|\varphi(e^{i\theta})|^p}{|1-e^{i(\alpha+n\theta)}|^p} d\theta$$

and hence

$$(1/2\pi) \int_0^{2\pi} \|f_\alpha\|_p^p d\alpha$$

$$= (1/4\pi^2) \int_0^{2\pi} |\varphi(e^{i\theta})|^p \int_0^{2\pi} |1-e^{i(\alpha+n\theta)}|^{-p} d\alpha d\theta$$

$$\leq c^p$$

where

$$c^p = (1/2\pi) \int_0^{2\pi} |1-e^{i(\alpha+n\theta)}|^{-p} d\alpha$$

is independent of n.

Hence for some α, $-\pi < \alpha \leq \pi$,

$$\|f_\alpha\|_p \leq c.$$

Now

$$\varphi(e^{i\theta}) - f_\alpha(e^{i\theta}) = (1-e^{i(\alpha+n\theta)})^{-1}\varphi(e^{i\theta})$$

and so

$$\varphi(e^{i\theta}) - f_\alpha(e^{i\theta}) = -e^{-i(\alpha+n\theta)}(1-e^{i(\alpha+n\theta})^{-1}\varphi(e^{i\theta})$$

$$= \bar{h}(e^{i\theta})$$

where $h \in H^p$ is given by

$$h(z) = -e^{i\alpha}/(1-e^{i\alpha}z^n) \sum_{k=0}^{2n} \bar{a}_{n-k}z^k.$$

Thus $\varphi - f_\alpha \in \overline{H}^p$ and

$$\|\varphi-f_\alpha\|_p^p \leq 1 + c^p.$$

Taking $C = (1+c^p)^{1/p}$, we have the desired conclusion. □

COROLLARY 3.11. J_p is weakly dense in H^p.

<u>Proof</u>. Suppose $\chi \in (H^p)^*$ and $\chi(J_p) = 0$. Consider the linear functional ψ on $H^p \oplus \overline{H}^p$ given by $\psi(f,g) = \chi(f) - \overline{\chi(\bar{g})}$. If $f \in J_p$

$$\psi(f,-f) = \chi(f) + \overline{\chi(\bar{f})} = 0.$$

Thus ψ factors to a linear functional ψ_1 on $(H^p \oplus \overline{H}^p)/E$ where $E = ((f,-f) : f \in J_p)$. By the theorem of Aleksandrov, this quotient is isomorphic to L_p and hence $\psi_1 = 0$. Thus $\psi = 0$ and $\chi = 0$ as required. □

We now turn to the case of invariant subspaces. We first observe that a closed subspace M of H^p invariant under multiplication by z is necessarily of the form SH^p where S is an inner function. This theorem is due to Beurling [1949] and Gamelin [1966].

The first example of a measure μ so that $S_\mu H^p$ is a PCWD-subspace was given by Duren, Romberg and Shields [1969]. Recently however Roberts [to appear] and Korenblum [1981] have succeeded in giving necessary and sufficient conditions on the measure μ for $S_\mu H^p$ to be PCWD.

We shall here describe a method of constructing examples based on Roberts's proof of the general necessary and sufficient condition. This approach uses the Corona Theorem and may be viewed as unnecessarily complicated for this purpose. However it both suggests the argument for the more general result and will help us to establish certain other properties of the quotient $H^p/S_\mu H^p$ later.

We first state the Corona Theorem of Carleson (see Duren, p. 202, Gamelin [1980], Koosis [1980]).

THEOREM 3.12. (Corona Theorem). For all $n \in \mathbf{N}$ there is a constant γ_n, so that if $0 < \delta < 1/2$, and $f_1,\dots,f_n \in H^\infty$ with $\|f_k\|_\infty \leqslant 1$ $(1 \leqslant k \leqslant n)$ and

$$\sum_{k=1}^{n} |f_k(z)| \geqslant \delta \qquad |z| < 1$$

then there exist $g_1,\dots,g_n \in H^\infty$ with $\|g_k\| \leqslant \delta^{-\gamma_n}$ $(1 \leqslant k \leqslant n)$ and

$$\sum_{k=1}^{n} g_k(z)f_k(z) = 1 \qquad |z| < 1.$$

We consider the set $M_1(\mathbf{T})$ of probability measures on $\mathbf{T}$ with the weak*-topology (induced by $C(\mathbf{T})$).

LEMMA 3.13. If $\mu_n \in M_1(\mathbf{T})$ and $\mu_n \to \mu$ weak* then $S_{\mu_n}(z) \to S_\mu(z)$ uniformly on compact subsets of D (where as usual

$$S_\nu(z) = \exp(-\int_0^{2\pi} (e^{it}+z)/(e^{it}-z)d\nu(t))).$$

Proof. For $|z| < 1$,

$$(e^{it}+z)/(e^{it}-z) \in C(\mathbf{T}).$$

Hence $S_{\mu_n}(z) \to S_\mu(z)$ pointwise on D. As $|S_{\mu_n}(z)| \leq 1$ for $z \in D$, (S_{μ_n}) is a normal family and the lemma follows. □

If $\mu \in M_1(\mathbf{T})$ is singular then it may be shown that $S_\mu^{-1} \notin H^\infty$ i.e.

$$\lim_{r \to 1} \inf_{-\pi<\theta\leq\pi} |S_\mu(re^{i\theta})| = 0.$$

However we have the following:

LEMMA 3.14. Suppose $r_n < 1$ and $\delta_n \downarrow 0$. Then there is a singular measure $\mu \in M_1(\mathbf{T})$ with

$$\inf_{-\pi<\theta\leq\pi} |S_\mu(r_n e^{i\theta})| \geq \delta_n$$

for infinitely many n.

Proof. Let E be a closed subset of T, with $m(E) = 0$ (here m is normalized Haar measure on the circle i.e. $dm = (2\pi)^{-1}d\theta$). We denote by $M_1(E)$ the set of probability measures supported on E, and by $\Lambda(E)$ the set of $\mu \in M_1(E)$ of the form

$$d\mu = \varphi \cdot dm$$

where $\varphi \in L_\infty(m)$.

We construct inductively a decreasing sequence of closed sets (E_k), an increasing sequence of integers (n_k) and a sequence $\mu_k \in \Lambda(E_k)$ so that

(3.11) Each E_k is a finite union of non-degenerate closed intervals

(3.12) $$\inf_{-\pi<\theta\leq\pi} |(S_{\mu_k}(r_{n_j} e^{i\theta})| > \delta_{n_j} \quad 1 \leq j \leq k.$$

Let us suppose $E_1,\ldots,E_k$, $n_1,\ldots,n_k$ and $\mu_1,\ldots,\mu_k$ have been chosen. Since $\mu_k \in \Lambda(E_k)$

$$\inf_{z\in D} |S_{\mu_k}(z)| > 0.$$

Choose $n_{k+1} > n_k$ so that

$$\inf_{-\pi<\theta\leq\pi} |S_{\mu_k}(r_{n_{k+1}} e^{i\theta})| > \delta_{n_{k+1}} .$$

Now we may approximate μ_k by a probability measure ν whose support is a finite subset of E_k, using Lemma 3.13, so that

$$\inf_{-\pi<\theta\leq\pi} |S_\nu(r_{n_j} e^{i\theta})| > \delta_{n_j} \qquad 1 \leq j \leq k+1.$$

Next choose $E_{k+1} \subset E_k$ to be a finite union of nondegenerate closed intervals so that $\text{supp}\ \nu \subset E_{k+1}$ and $m(E_{k+1}) < (k+1)^{-1}$. Finally approximate ν by $\mu_{k+1} \in \Lambda(E_{k+1})$ so that

$$\inf_{-\pi<\theta\leq\pi} |S_{\mu_{k+1}}(r_{n_j} e^{i\theta})| > \delta_{n_j} \qquad 1 \leq j \leq k+1.$$

To complete the proof of the lemma let μ be any weak*-cluster point of the sequence (μ_k). Then for each k, if $\varphi \in C(\mathbf{T})$ and $\varphi | E_k = 0$ then

$$\int \varphi d\mu = 0.$$

Hence $\text{supp}\ \mu \subset \bigcap_{k=1}^{\infty} E_k$, i.e. μ is singular. On the other hand

$$\inf_{-\pi<\theta\leq\pi} |S_\mu(r_{n_j} e^{i\theta})| \geq \delta_{n_j} \qquad j = 1,2,\ldots \ . \ \square$$

THEOREM 3.15. There is a singular probability measure μ so that $S_\mu H^p$ is a PCWD subspace of H^p.

<u>Proof</u>. Let us denote by $h_n \in H^p$ the function

$$h_n(z) = z^n \qquad z \in D.$$

Then $\|h_n\|_p = 1$, but

$$\|h_n\|_{p,1} = 2\int_0^1 r^{n+1}(1-r)^{1/p-2}\,dr,$$

and so $\|h_n\|_{p,1} \to 0$, by the dominated convergence theorem. Choose $\beta_n = \|h_n\|_{p,1}^{-1/2}$ so that $\|\beta_n h_n\|_{p,1} \to 0$ but $\beta_n \to \infty$. Let $\delta_n = \beta_n^{-1/\gamma_2}$ where γ_2 is chosen as in the Corona Theorem (3.12). Let $r_n = \delta_n^{1/n}$.

Now since $\delta_n \to 0$ there is a singular probability measure μ so that

$$(3.13) \qquad \inf_{-\pi<\theta\leqslant\pi} |S_\mu(r_n e^{i\theta})| \geqslant \delta_n$$

for infinitely many n. Let M be the collection of $n \in \mathbf{N}$ so that (3.13) holds.

Then for $n \in M$, since S_μ^{-1} is analytic for $|z| < 1$,

$$|S_\mu(z)| \geqslant \delta_n \qquad |z| \leqslant \gamma_n$$

while if $|z| > \gamma_n$

$$|h_n(z)| > r_n^n = \delta_n.$$

Hence

$$|S_\mu(z)| + |h_n(z)| \geqslant \delta_n \qquad |z| < 1,$$

and by Theorem 3.12, there exist $f_n, g_n \in H_\infty$ with $\|f_n\|_\infty$,

$\|g_n\|_\infty \leq \delta_n^{-\gamma_2} = \beta_n$ and

$$f_n S_\mu + g_n h_n = 1.$$

Suppose $\chi \in (H^p)^*$ and $\chi(S_\mu H^p) = 0$. Then for some constant $C < \infty$

$$|\chi(\varphi)| \leq C\|\varphi\|_{p,1} \qquad \varphi \in H^p.$$

Now if $\varphi \in H_\infty$ and $n \in M$

$$\varphi = \varphi f_n S_\mu + \varphi g_n h_n$$

and so

$$\chi(\varphi) = \chi(\varphi g_n h_n).$$

However

$$\begin{aligned}\|\varphi g_n h_n\|_{p,1} &\leq \|\varphi\|_\infty \|g_n\|_\infty \|h_n\|_{p,1} \\ &\leq \|\varphi\|_\infty \|h_n\|_{p,1}^{1/2} \\ &\to 0\end{aligned}$$

as $n \to \infty$. Thus $\chi(\varphi) = 0$ and so $\chi(H^p) = 0$. □

A more delicate handling of this argument yields the general classification:

THEOREM 3.16. (Roberts-Korenblum [to appear],[1981]). In order that $S_\mu H^p$ be a PCWD subspace of H^p it is necessary and sufficient that $\mu(C) = 0$ for every Carleson thin set C.

(A subset C of $\mathbf{T}$ is a Carleson thin set if it has measure zero and $\mathbf{T} \setminus C$ is the union of at most countably many arcs $\{I_n : n = 1,2,...\}$ so that

$$\sum_{n=1}^{\infty} m(I_n)\log (1/(m(I_n)) < \infty .)$$

CHAPTER 4
THE HAHN-BANACH EXTENSION PROPERTY

1. Introduction.

We have already seen that various forms of the Hahn-Banach Theorem fail in nonlocally convex F-spaces. It is natural to ask whether any of the Hahn-Banach formulations ever hold in a nonlocally convex F-space. The moral of this chapter is that, generally, such formulations imply local convexity.

Suppose that X is an F-space and M is a closed subspace of X. M is said to have the Hahn-Banach Extension Property (HBEP) if whenever $\varphi \in M^*$ there exists $x^* \in X^*$ such that $\varphi(x) = x^*(x)$ for every $x \in M$. M is weakly closed if whenever $x \notin M$ there exists $x^* \in X^*$ such that $x^*(M) = \{0\}$ and $x^*(x) \neq 0$. Recall from Chapter 2 that a closed subspace M is proper closed weakly dense (PCWD) if $x^* \in X^*$ and $x^*(M) = \{0\} \implies x^* \equiv \{0\}$. The notions of weakly closed and weakly dense can, of course, be defined in terms of the (possibly non-Hausdorff) weak topology on X. Observe that X has the point separation property if and only if the weak topology is Hausdorff. The proof of the following lemma is straightforward.

LEMMA 4.1. If X is an F-space, then

(1) X has the point separation property if and only if every finite dimensional subspace of X has HBEP

(2) M is weakly closed in X if and only if X/M has the point separation property

(3) M is PCWD in X if and only if X/M is a nontrivial space with trivial dual.

One might expect that there is some connection between a closed subspace having HBEP or being weakly closed. Unfortunately there is no such relationship for a single subspace, as we shall see in the following two examples.

The examples are constructed by considering a sequence (x_n) in ℓ_p whose supports are disjoint, such that $\|x_n\|_p = 1$ but $\|x_n\|_1 \leq 2^{-n}$.

EXAMPLE A. weakly closed =/=> HBEP.

Let $M = \overline{\text{span}}(x_n)$. Since (x_n) is a basic sequence equivalent to the ℓ_p basis, there is a linear functional φ in M^* such that $\varphi(x_n) = 1$ for all n. Since $\|x_n\|_1 \to 0$, φ cannot be extended to ℓ_p.

EXAMPLE B HBEP =/=> weakly closed.

Let $z_1 = x_1$ and for $n \geq 2$ let $z_n = x_n + e_{n-1}/2$ where (e_n) is the standard basis of ℓ_p. Let $E = \text{span}(z_n)$ for $n \geq 2$. A routine calculation shows that $(z_n : n \geq 1)$ is equivalent to the standard basis of ℓ_p. Hence E is a proper closed subspace of ℓ_p. However, a similar calculation in the ℓ_1 norm shows that $(x_n : n \geq 2)$ is a basis of ℓ_1 equivalent to the unit vector basis. Hence E is weakly dense in ℓ_p. If $\varphi \in E^*$ then $(\varphi(z_n))$ is bounded and therefore φ extends to an element of $\ell_1^* = \ell_p^*$.

Despite the fact that there is no connection between the two notions we now show that there is a global equivalence.

DEFINITION. An F-space X has the HBEP if every closed subspace M of X has HBEP.

LEMMA 4.2. X has HBEP if and only if every closed subspace of X is weakly closed.

Proof. Suppose X has HBEP. If M is a closed subspace, it is clear that X/M has HBEP. But then X/M has the point separation property. Now suppose every closed subspace of X is weakly closed. Let M be a closed subspace of X and let $\varphi \in M^* \sim \{0\}$. Then $N = \{x \in M : \varphi(x) = 0\}$ is a closed subspace of X. Now $\varphi(x) \neq 0$ for some $x \in M$, so $M = \mathbb{R}x + N$. Since N is weakly closed there exists $\psi \in X^*$, such that $\psi(N) = \{0\}$ and $\psi(x) = \varphi(x)$. Thus ψ extends φ to X. □

We have seen in Chapters 2 and 3 that the classical spaces ℓ_p and H^p $(0 < p < 1)$ possess PCWD subspaces. By Lemma 4.2 this in turn implies that they do not have the HBEP. Motivated by these and other examples, Duren, Romberg and Shields [1969] formulated the following problems:

PROBLEM 4.1. If X is an F-space with HBEP, is X locally convex?

In fact a question related to Problem 4.1 was first raised by M. Henriksen in the 1950's. He asked what assumptions on a closed subspace M of a topological vector space with the point separation property would force the quotient X/M to have the point separation property. In response, Klee [1956] constructed a linear metric space X (not complete) with the point separation property such that for a particular closed subspace M, X/M failed the separation property. Later, Peck [1965] produced a closed subspace M of ℓ_p such that ℓ_p/M fails the separation property (i.e M is not weakly closed). Of course, we have seen in Chapter 1 that ℓ_p has a PCWD subspace, a result due to Stiles [1970] and Shapiro [1969].

In problem 4.1, if we relax the metrizability of X then the question becomes trivial. Shuchat observed that if X is any vector space of uncountable algebraic dimension with its strongest vector topology then X has HBEP and is not locally convex (if X has merely countable dimension, it is locally convex). (See Duren, Romberg, and Shields [1969, p. 59].) Also, Gregory and Shapiro [1970] showed that is always possible to interpolate a nonlocally convex topology between the weak and strong topologies of an infinite-dimensional Banach space, to provide another example.

The first results on Problem 4.1 were obtained by Shapiro [1970] who showed that the answer is affirmative if X has a basis. Kalton [1974] showed that the answer to Problem 4.1 is in general affirmative, by developing basic sequence techniques in general F-spaces. In this chapter we give an exposition of this result, following a later simplification by Drewnowski [1977] of some of the ideas in Kalton's 1974 paper.

A second problem related to 4.1 was also raised by Duren, Romberg and Shields:

PROBLEM 4.2. Let X be an F-space which is not locally convex. Must X possess a PCWD subspace?

In this chapter we shall see that the answer to Problem 4.2 is affirmative if we assume that X is separable and has the point separation property; this result is due to Kalton [1978e]. It is clearly also affirmative if X has trivial dual. However, in the next chapter we shall see that in general the answer to Problem 4.1 is negative (Roberts [1977b]).

In the last section we shall discuss the properties of the containing Banach space of a nonlocally convex locally bounded F-space. We have seen that the containing Banach space of ℓ_p is ℓ_1 for $0 < p < 1$. As noted in Chapter 3, a result of Lindenstrauss and Pelczynski [1971] shows that the containing Banach space, B_p, of H^p, for $0 < p < 1$, is also isomorphic to ℓ_1. This coincidence prompted Shapiro to ask whether one could have the containing Banach space isomorphic to, for example, ℓ_2. It turns out that this is impossible and containing Banach spaces must always be "ℓ_1-like" in some sense (Kalton (1978b)).

2. Basic Sequences and the Hahn Banach Extension Property

DEFINITION. Let E be a linear space and suppose that σ and γ are two vector topologies on E. We say that σ is γ-*polar* if σ has a base of γ-closed neighborhoods of the origin.

A simple example of this is obtained by letting E be any normed linear space with σ and γ the norm and weak topologies respectively.

LEMMA 4.3. Let X be a metrizable linear space and suppose that γ is a vector topology on X so that the metric topology is γ-polar. Then the metric topology is given by an F-norm $\|\cdot\|$ such that for every $x \in X$

$$\|x\| = \sup\{\lambda(x) : \lambda \in \Lambda\}$$

where Λ is a collection of γ-continuous F-seminorms on X. In particular, $\|\cdot\|$ is lower semicontinuous with respect to γ.

Proof. Let Γ be the collection of *all* γ-continuous F-seminorms on X and let $\|\cdot\|_0$ be *any* F-norm defining the metric topology of X. For each $\eta \in \Gamma$ define $\lambda_\eta(x)$ for each $x \in X$ by

$$\lambda_\eta(x) = \inf\{\|y\|_0 + \eta(x-y) : y \in X\}.$$

Let $\Lambda = \{\lambda_\eta : \eta \in \Gamma\}$ and define $\|x\| = \sup_{\lambda \in \Lambda} \lambda(x)$. Clearly

$\|x\| \leq \|x\|_0$. Now suppose that V is a γ-closed 0-neighborhood. For some $\epsilon > 0$, $\|x\|_0 < \epsilon$ implies $x \in V$. We claim that if $\|x\| < \epsilon$ then $x \in V$. Indeed, if $\|x\| < \epsilon$, then for any γ-neighborhood U of 0, there exists $\eta \in \Gamma$ so that $\eta(u) < \epsilon$ implies $u \in U$. Since $\lambda_\eta(x) < \epsilon$ there exists y so that $\|y\|_0 < \epsilon$ and $\eta(x-y) < \epsilon$. Hence $x \in V+U$. But this is true for every γ-neighborhood U of 0. Thus since V is γ-closed, $x \in V$. □

LEMMA 4.4. Suppose, under the same hypotheses as Lemma 4.5, (x_n) is a sequence converging to 0 in γ, which is Cauchy in the original topology. Then $x_n \to 0$ in the original topology.

Proof. Let $\|\cdot\|$ be the F-norm constructed in Lemma 4.3. If $\epsilon > 0$ there exists N such that if $m,n \geq N$, then $\|x_m - x_n\| \leq \epsilon$. For any fixed $n \geq N$, $x_n - x_m \to x_n$ with respect to γ. Since $\|\cdot\|$ is lower semicontinuous with respect to γ, $\|x_n\| \leq \epsilon$. Thus $\|x_n\| \to 0$. □

LEMMA 4.5. Let X be a separable F-normed space and let γ be a vector topology on X so that the original topology is γ-polar. Then there is a metrizable vector topology γ' on X such that $\gamma' \leq \gamma$ and the original topology is γ'-polar.

Proof. Since X is separable there exists a sequence (F_n) of finite subsets of X such that $\bigcup_{n=1}^{\infty} F_n$ is dense. We can suppose that X is F-normed as in Lemma 4.3. Choose Λ_n a finite set in Λ such that for each $x \in F_n$

$$\max_{\lambda \in \Lambda_n} \lambda(x) \leq \|x\| \leq \max_{\lambda \in \Lambda_n} \lambda(x) + 1/n.$$

Now let $\Lambda_\infty = \bigcup_{n=1}^{\infty} \Lambda_n$. Then for $x \in \bigcup_{n=1}^{\infty} F_n$ (and hence for any $x \in X$)

$$\|x\| = \sup_{\lambda \in \Lambda_\infty} \lambda(x).$$

Since Λ_∞ is countable, it generates a metrizable vector topology $\gamma' \leq \gamma$ and clearly the original topology is γ'-polar. □

We now come to a basic result due to Drewnowski [1977] which requires the notion of an M-basic sequence.

DEFINITION. A sequence (x_n) in an F-normed space X is M-<u>basic</u> if there is a sequence (x_n^*) of continuous linear functionals defined on the closed linear span X_0 of $\{x_n : n \in N\}$ such that

(a) $x_i^*(x_j) = \delta_{ij}$

(b) if (u_n) is a Cauchy sequence in X_0 and $x_i^*(u_n) \to 0$ for every i, then $u_n \to 0$.

If, in addition, $\{x_n^* : n \in N\}$ is equicontinuous, we say that (x_n) is <u>strongly regular</u>.

Note that if X is complete, then (b) is equivalent to the statement that $\{x_n^* : n \in N\}$ is total. We shall, however, require the notion for incomplete spaces.

THEOREM 4.6. Let $(X, \|\cdot\|)$ be an F-normed space and let ρ denote the F-norm topology on X. Suppose γ is a weaker vector topology on X such that ρ is γ-polar. Further, suppose that (x_n) is a sequence in X such that $x_1 \neq 0$, $x_n \to 0$ in γ but $\|x_n\| \nrightarrow 0$. Then there is a subsequence (z_n) of (x_n) with $z_1 = x_1$ such that (z_n) is M-basic and strongly regular in $(X, \|\cdot\|)$.

<u>Proof</u>. Without loss of generality we may assume that X is separable and that γ is metrizable. We may further suppose that $\| \ \|$ is γ-lower semicontinuous and $| \ |$ is an F-norm defining the topology of γ. By passing to a subsequence (but not removing x_1) we can assume that

$$\sum_{n=1}^{\infty} |x_n| < \infty$$

while $\|x_n\| \geq \epsilon > 0$ for all n.

The sequence (x_n) cannot have any Cauchy subsequences by Lemma 4.4. Consequently $\{x_n : n \in N\}$ cannot be totally bounded in (X, ρ). Hence there exists $\delta > 0$ with $\|x_1\| > \delta$ so that if K is a ρ-compact subset of X there exists $n \in N$ such that for all $u \in K$

$$\|x_n - u\| \geq \delta.$$

If $n \in N$, then the set

$$\{\sum_{i=1}^{n} a_i x_i : |a_i| \leq 1, \; 1 \leq i \leq n\}$$

is compact and so by induction we can find a subsequence (y_n) of (x_n) with $y_1 = x_1$ and so that for every n, if $|t_n| = 1$ and $|t_i| \leq 1$ for $1 \leq i < n$,

$$\|\sum_{i=1}^{n} t_i y_i\| \geq \delta.$$

Next we claim that for any $n \in N$ there exists $m > n$ so that whenever $|t_i| \leq 1$ for $1 \leq i < n$ and $m \leq i \leq r$ and $|t_n| = 1$ then

$$\|\sum_{i=1}^{n} t_i y_i + \sum_{i=m}^{r} t_i y_i\| \geq \delta/2.$$

Suppose not. Then there exist

$$w_k = \sum_{i=1}^{n} t_{ik} y_i$$

$$v_k = \sum_{i=m(k)}^{r(k)} t_{ik} y_i$$

where $|t_{nk}| = 1$ and $|t_{ik}| \leq 1$ for all i and k, $m(k) \to \infty$ and

$$\|w_k + v_k\| < \delta/2.$$

Since $\sum_{i=1}^{\infty} |y_i| \leq \sum_{i=1}^{\infty} |x_i| < \infty$, we conclude that

$|v_k| \to 0$. By an elementary compactness argument (w_k) has a cluster point w (in both ρ and γ) where

$$w = \sum_{i=1}^{n} t_i^* y_i$$

with $|t_i^*| \leq 1$ for $1 \leq i < n$ and $|t_n^*| = 1$. Since $\|\cdot\|$ is γ-lower semicontinuous

$$\|w\| \leq \sup_k \|w_k + v_k\| \leq \delta/2,$$

contradicting the choice of (y_n).

Now by induction we can select a subsequence (z_n) of (y_n) with $z_1 = y_1 = x_1$ so that

$$\left\| \sum_{i=1}^{n} t_i z_i \right\| > \delta/2.$$

whenever $\max\{|t_i| : 1 \leq i \leq n\} \geq 1$. A trivial consequence of this is that (z_n) is linearly independent. Let (z_n^*) be the corresponding biorthogonal sequence on the linear span of $\{z_n : n \in \mathbf{N}\}$. But then for any $n \in \mathbf{N}$

$$|z_n^*(x)| \leq 1$$

if $\|x\| \leq \delta/2$. Hence (z_n^*) is equicontinuous and each z_n^* may be extended to the closed linear span E of $\{z_n : n \in N\}$. Finally suppose (u_n) in E is a Cauchy sequence with $z_i^*(u_n) \to 0$ for every i. Pick (v_n) in the linear span of (z_n) so that for every n, $\|u_n - v_n\| \leq 1/n$. Then (v_n) is also Cauchy. By the equicontinuity of (z_n^*) there exists a constant

$K > 0$ so that for every v_n

$$\sup_i |z_i^*(v_n)| \leq K.$$

Since

$$v_n = \sum_{i=1}^{\infty} z_i^*(v_n)z_i$$

$z_i^*(v_n) \to 0$ and $\sum_{n=1}^{\infty} |z_n| < \infty$, it is easy to see that $v_n \to 0(\gamma)$. Hence by Lemma 4.4 $v_n \to 0(\rho)$ and thus $u_n \to 0(\rho)$. □

REMARKS. With more work it is possible to select a basic subsequence (Kalton, 1974) and to eliminate the metrizability assumption. Theorem 4.6 is essentially the key to the solution of Problem 4.1, since it enables us to construct continuous linear functionals on subspaces of an F-space. One would hope to proceed by allowing ρ to be the original topology on X and γ to be the <u>Mackey topology</u> on X, i.e., the metrizable topology obtained by taking all convex ρ-neighborhoods of the origin as the neighborhood base at the origin. The problem is that, in general, γ need not be ρ-polar. We now address this problem.

THEOREM 4.7. Suppose X is an F-space and α is a Hausdorff vector topology on X weaker than the original topology. Suppose (x_n) is a sequence in X with $x_1 \neq 0$ such that $x_n \not\to 0$ but $x_n \to 0(\alpha)$. Then (x_n) has a subsequence (z_n) which is M-basic and strongly regular.

<u>Proof</u>. Let γ be the largest vector topology on X such that $x_n \to 0(\gamma)$ but γ is weaker than the original topology. Thus an

F-seminorm η is γ-continuous if and only if η is continuous in the original topology and $\eta(x_n) \to 0$. Since $\alpha \leq \gamma$, γ is Hausdorff.

Now let ρ be the topology on X generated by the base at zero consisting of all γ-closed 0-neighborhoods. Thus if $\{V_n : n \in N\}$ is a base at zero for the original topology then the sets $\overline{V}_n$ (closure in γ) form a base for ρ. Clearly ρ is metrizable, $\gamma \leq \rho$, ρ is γ-polar, and the original topology is at least as strong as ρ.

Suppose $x_n \to 0(\rho)$; then $\rho = \gamma$. Consider the identity map $i : X \to (X,\gamma)$. This map is continuous, and if V is a 0-neighborhood in X, then $\overline{i(V)}$ is a ρ-neighborhood and hence a γ-neighborhood. But then i is an isomorphism and γ is the original toplogy (Theorem 1.4). This is a contradiction since $x_n \to 0(\gamma)$. Hence $x_n \not\to 0(\rho)$.

Now by Theorem 4.6 (x_n) has a subsequence (z_n) which is M-basic and strongly regular in (X,ρ). Let E be the linear span of (z_n).

We claim that (z_n) is also M-basic and strongly regular in X with its original topology. Clearly the biorthogonal functionals (z_n^*) are also equicontinuous on E in the original topology and may be extended to an equicontinuous collection on the closed linear span X_0 of E. Suppose (u_n) in X_0 is a Cauchy sequence. Then (u_n) is also ρ-Cauchy. If $z_i^*(u_n) \to 0$ for every i then $u_n \to 0(\rho)$. Since X is complete, (u_n) converges in X. Since ρ is Hausdorff we may conclude that $\|u_n\| \to 0$. □

We now come to the main result of this section.

THEOREM 4.8. If X is an F-space with HBEP, then X is locally convex.

<u>Proof</u>. Let us denote by μ the Mackey topology on X, i.e., the finest locally convex topology weaker than the original topology. Since μ is generated by the sets $\text{co}\, V_n$ where $\{V_n : n \in N\}$ is a base at zero for the original topology, μ is metrizable. Also μ is finer than the weak topology of X, i.e. every linear functional which is continuous on X is also continuous on (X,μ). By Lemma 4.2 every closed subspace of X is μ-closed.

Since μ is metrizable it suffices to show that if $w_n \to 0(\mu)$ then $\{w_n : n \in N\}$ is bounded or equivalently $c_n w_n \to 0$ whenever $c_n \to 0$. Pick any $u \in X$ with $u \neq 0$. Define $x_n = c_n(u+w_n)$. Then $x_n \to 0(\mu)$. Suppose $\|x_n\| \not\to 0$. Then there is a subsequence (z_n) of (x_n) which is M-basic. Now let L_n be the closed linear span of $\{z_k : k \geq n\}$. If $v \in \bigcap_{n=1}^{\infty} L_n$ then for every i, $z_i^*(v) = 0$ and hence $v = 0$, i.e., $\bigcap_{n=1}^{\infty} L_n = \{0\}$. But if $z_n = x_{m_n}$, then $c_{m_n}^{-1} z_n \to u(\mu)$. Each L_n is μ-closed so that $u \in \bigcap_{n=1}^{\infty} L_n$. Thus $u = 0$ contrary to assumption. Therefore $\|x_n\| \to 0$ and hence $\| c_n w_n \| \to 0$ whenever $c_n \to 0$. □

An open question related to Theorem 4.8 is the following:

PROBLEM 4.3. Is there a nonlocally convex F-space with separating dual so that every weakly closed subspace has HBEP?

We close this section with some remarks on the existence of basic sequences. As we pointed out after Theorem 4.6, the methods of this section can be employed to construct basic sequences under quite general conditions. We call an F-space minimal if it admits no Hausdorff vector topology strictly weaker than the original topology. The most general form of Theorem 4.6 is then:

THEOREM 4.9. [Kalton-Shapiro [1976]]. Let X be a non-minimal F-space. Then X contains a basic sequence.

The only known example of a minimal space is the space ω of all sequences, which of course is locally convex and has a basis (Kalton [1974], Drewnowski [1977a]). It is obviously of interest to answer:

PROBLEM 4.4. Does every F-space contain a basic sequence?

PROBLEM 4.5. Does there exist a non-locally convex minimal space?

Drewnowski [1979] has studied the stronger notion of a quotient-minimal space; X is quotient-minimal if every quotient of X is also minimal. Alternatively X is quotient-minimal if every operator $T : X \to Y$ has closed range or if there are no non-trivial weaker vector topologies on X. The space ω is quotient-minimal.

PROBLEM 4.6. Does there exist a non-locally convex quotient-minimal space?

Finally we note another possible monster. Let us say X is atomic if every proper closed subspace of X is finite-dimensional. Of course an atomic space is separable.

PROBLEM 4.7. Does there exist an atomic F-space?

It is quite clear that any atomic space must be minimal (by Theorem 4.9) and indeed it must be quotient-minimal. There is a partial converse to this.

THEOREM 4.10. Let X be any locally bounded F-space. Then either

(a) X contains an infinite-dimensional atomic subspace

or

(b) X has a finite-dimensional subspace F so that X/F is not minimal.

Proof. Let U be the unit ball of X. Let $\mathcal{L}$ be any maximal collection of infinite-dimensional closed subspaces of X so that

if $L_1, \ldots, L_n \in \mathcal{L}$ then $L_1 \cap \ldots \cap L_n \in \mathcal{L}$. Let $F = \cap\mathcal{L}$. If $\dim F = \infty$ then F is atomic.

Alternatively let $\dim F$ be finite. Then we can define a vector topology ρ on X with a base of neighborhoods of 0 of the form $L + \epsilon U$ where $L \in \mathcal{L}$ and $\epsilon > 0$. Since every neighborhood of 0 contains a line ρ is a strictly weaker vector topology on X. Also one can see that $\bigcap_{L\in\mathcal{L}} \bigcap_{\epsilon>0} (L+\epsilon U) = F$, so that ρ factors to a Hausdorff topology on X/F which is strictly weaker than the locally bounded topology. □

Although we do not know whether an atomic space exists, in Chapter 7 we shall construct a rigid F-space: an F-space whose only endomorphisms are scalar multiples of the identity. An atomic space must necessarily be rigid; see the discussion in Chapter 7.

3. The Construction of PCWD Subspaces

Before proceeding to a general result on the existence of PCWD spaces we shall need a lemma on compact operators. We shall consider such operators in greater detail in Chapter 7.

DEFINITION. Let X and Y be F-spaces. A linear operator $K : X \to Y$ is <u>compact</u> if there exists a 0-neighborhood U in X such that $K(U)$ is relatively compact.

LEMMA 4.11. Suppose X and Y are F-spaces and $A : X \to Y$ is an isomorphism onto its range. If $K : X \to Y$ is compact, then $A+K$ has closed range.

<u>Proof</u>. Let U be a 0-neighborhood in X such that $K(U)$ is relatively compact in Y. Also let $N = \{x \in X : (A+K)(x) = 0\}$ and let $q : X \to X/N$ be the natural quotient map. Then we can define S by $A+K = Sq$ and we need only show that S is a linear isomorphism.

If S is not a linear isomorphism there is a 0-neighborhood V in X with $V \subset (1/2)U$ such that for some sequence $u_n \in X/N$ with $u_n \notin q(V)$ for all n we have $Su_n \to 0$. We can select a sequence (a_n) of scalars so that $0 < a_n \leq 1$, $a_n u_n \notin q(V)$ for all n, but $a_n u_n \in q(U)$ since $q(V) \subset (1/2)q(U)$. Let $a_n u_n = q(x_n)$ where $x_n \in U$. Then $(K(x_n))$ has a cluster point $y \in Y$. However $Sqx_n = (A+K)(x_n) \to 0$, so that $-y$ is a cluster point of $(A(x_n))$. Thus there is a subsequence (w_n) of (x_n) with $K(w_n) \to y$ and $A(w_n) \to -y$. Since A is an isomorphism, $w_n \to w$ where $A(w) = -y$ and therefore $K(w) = y$. Hence $q(w) = 0$. But then 0 is a cluster point of $(a_n u_n)$ contrary to assumption. □

THEOREM 4.12. [Kalton [1978c]]: Let X be a separable F-space with separating dual. If X is not locally convex then X contains a PCWD subspace.

Proof. Since X is not locally convex, the Mackey topology μ on X is strictly weaker than the original topology. Hence by Theorem 4.7 there is an M-basic sequence (u_n) which is strongly regular but such that $u_n \to 0(\mu)$. Let $(u_n{}^*)$ denote the biorthogonal functionals defined on the closed linear span E of (u_n). Also let E_0 be the closed linear span of $\{u_{2n} : n \in N\}$. Since X is separable there exists a dense sequence (v_n) in X. Choose ϵ_n with $0 < \epsilon_n < 1$ such that

$$\|\epsilon_n v_n\| \leq 1/2^n \quad \text{for all } n.$$

Now $u_n \to 0(\mu)$ and μ is metrizable, so there exists a subsequence $(u_{2\ell(n)})$ with $\epsilon_n{}^{-1} u_{2\ell(n)} \to 0(\mu)$. Define $K : E_0 \to X$ by

$$K(x) = \sum_{n=1}^{\infty} \epsilon_n u_{2\ell(n)}{}^*(x) v_n.$$

The set $U = \{x \in E_0 : u_{2\ell(n)}{}^*(x) \leq 1 \text{ for every } n\}$ is a 0-neighborhood in E_0 which is mapped into a compact set since

$$\sum_{n=1}^{\infty} \|\epsilon_n v_n\| \leq \sum_{n=1}^{\infty} 1/2^n = 1.$$

Hence K is a compact operator.

Let $J : E_0 \to X$ be the inclusion map and let $N = (J+K)(E_0)$. We will show that N is a PCWD subspace.

N is closed by Lemma 4.11. We claim that N is proper. Indeed if $q : X \to X/E_0$ is the quotient map then

$q(J+K) = qK$ is compact. Since $\dim X/E_0 = \infty$, $q(J+K)$ cannot be surjective and hence $J+K$ cannot be surjective.

Finally we claim that N is weakly dense. Suppose $\varphi \in X^*$ and $\varphi(N) = \{0\}$. For any $x \in X$ there is a sequence $v_{n_k} \to x$. Now $\epsilon_{n_k}^{-1}(u_{2\ell(n_k)}+\epsilon_{n_k}v_{n_k}) = (J+K)(\epsilon_{n_k}^{-1}u_{2\ell(n_k)}) \in N$. Hence $\varphi(v_{n_k}) = -\varphi(\epsilon_{n_k}^{-1}u_{2\ell(n_k)})$. Since $\epsilon_{n_k}^{-1}u_{2\ell(n_k)} \to 0(\mu)$, $\varphi(x) = 0$. □

We shall see in Chapter 5 that a separating dual is necessary in Theorem 4.12. However, the following is an open question:

PROBLEM 4.8. Can the separability assumption be removed from Theorem 4.12?

4. Containing Banach Spaces

If X is a locally bounded F-space with separating dual then its Mackey topology is normable and when completed yields a Banach space, i.e., the containing Banach space $\hat{X}$ of X. As noted in the introduction, for many standard examples, $\hat{X}$ is isomorphic to ℓ_1.

We start out by giving an example due to A. Pelczynski (see Kalton and Shapiro [1976]) which shows that $\hat{X}$ can be a reflexive Banach space without X being locally convex.

Let (E_n) be a sequence of finite dimensional quasi-normed spaces where the quasi-norms satisfy

$$\|x+y\| \leq C(\|x\|+\|y\|) \quad \text{for all } x,y \in E_n$$

for some constant C independent of n. Form the space $\ell_2(E_n)$ of sequences $x = (x_n)$ with $x_n \in E_n$ such that

$$\|x\| = (\sum_{n=1}^{\infty} \|x_n\|^2)^{1/2} < \infty.$$

Then $\ell_2(E_n)$ is also a quasi-normed F-space. It is easily shown that the dual of $\ell_2(E_n)$ is the Banach space $\ell_2(E_n^*)$ of sequences (x_n^*) with $x_n^* \in E_n^*$ and

$$\|x^*\| = (\sum_{n=1}^{\infty} \|x_n^*\|^2)^{1/2} < \infty.$$

It is then easy to verify that the containing Banach space norm on $\ell_2(E_n)$ is given by

$$\|x\|_0 = (\sum_{n=1}^{\infty} \|x_n\|_0^2)^{1/2}$$

where each $\|x_n\|_0$ is the containing Banach space norm of $x_n \in E_n$. Hence $\hat{X} = \ell_2(\hat{E}_n)$.

Note that $\hat{X}$ is a reflexive Banach space. Now we let $E_n = \ell_p^{(n)}$ for a fixed p, $0 < p < 1$. Also there exists $x_n \in E_n$ such that $\|x_n\| = \|x_n\|_p = 1$ but $\|x_n\|_0 = \|x_n\|_1 \to 0$. Thus for this choice of E_n, $\ell_2(E_n)$ is not locally convex.

However, in spite of this example, there are some restrictions on the spaces $\hat{X}$ that can arise from a nonlocally convex space X. Recall that a Banach space Y is B-<u>convex</u> (Giesy [1966], Beck [1962]) if $b_n/n \to 0$ where

$$b_n = \sup_{\|y_i\| \le 1} \inf_{\epsilon_i = \pm 1} \|\epsilon_1 y_1 + \ldots + \epsilon_n y_n\|.$$

Y fails to be B-convex if and only if ℓ_1 <u>is finitely representable in</u> Y, i.e., for any $\epsilon > 0$ and $n \in N$ there is a linear isomorphism $T : \ell_1^{(n)} \to E$ where E is a subspace of Y and $\|T\|\|T^{-1}\| \leq 1 + \epsilon$. It is easy to show that $b_m b_n \geq b_{mn}$ for every $m,n \in N$ and to deduce that if X is B-convex we have an inequality of the form

$$b_n \leq Cn^{1-\alpha} \quad \text{for all } n \in N$$

where $\alpha > 0$.

THEOREM 4.13. Let X be a locally bounded F-space and let Y be a B-convex Banach space. Suppose $T : X \to Y$ is a continuous linear operator and that $\overline{\text{co}}\, T(B)$ is a 0-neighborhood (B denotes the unit ball in X). Then T is an open mapping.

<u>Proof</u>. Let $\|\cdot\|$ be a quasi-norm on X so that for some p, $0 < p < 1$, $\|\cdot\|^p$ is an F-norm. Without ambiguity we let $\|\cdot\|$ also denote the norm on Y. We shall suppose that $\|T\| = 1$ and that $\delta > 0$ is chosen so that if $\|y\| < \delta$ then $y \in \overline{\text{co}}\, T(B)$. It will suffice to show the existence of a constant $M < \infty$ so that if $\|y\| \leq 1$, there exists $x \in X$ with $\|x\| \leq M$ and $\|T(x)-y\| < 1/2$. For if we can do this we may then choose x_n by induction so that $\|x_n\| \leq M2^{-n}$ for $n \geq 0$ and

$$\|T(x_0+\ldots+x_n)-y\| \leq 1/2^{n+1}.$$

Then $T(\sum_{i=0}^{\infty} x_i) = y$ (the series $\sum_{i=0}^{\infty} x_i$ converges since $\sum_{i=0}^{\infty} \|x_i\|^p < \infty$).

Now let

$$V_n = \{(1/k)(T(x_1)+\ldots+T(x_k)) : \|x_i\| \leq 1,\ k \leq n\}.$$

Then if $y \in V_n$

$$y = (1/k)(T(x_1)+\ldots+T(x_k))$$

where $1 \leq k \leq n$ and $\|x_i\| \leq 1$. We can choose $\epsilon_i = \pm 1$ so that

$$\|\sum_{i=1}^{k} \epsilon_i T(x_i)\| \leq b_k \leq Ck^{1-\alpha}$$

and we may suppose that $\{i : \epsilon_i = -1\}$ has at most $k/2$ members (by using $-\epsilon_i$ instead of ϵ_i for each i, if necessary). Let

$$y_1 = (2/k) \sum_{\epsilon_i=-1} T(x_i).$$

Then $y_1 \in V_{[n/2]}$ and

$$\|y-y_1\| \leq Ck^{-\alpha}.$$

In particular, if $k > n/2$

$$\|y-y_1\| \leq C_1 n^{-\alpha}$$

where $C_1 = 2^{\alpha}C$. Thus in general if $y \in V_n$ then

$$d(y,V_{[n/2]}) \leq C_1 n^{-\alpha}.$$

Hence if $y \in V_{2^{m+n}}$ then

$$d(y,V_{2^m}) \leq C_1 \sum_{j=m+1}^{m+n} 2^{-j\alpha} \leq C_2 2^{-m\alpha}$$

where $\underline{C_2 = C_1} \sum_{j=1}^{\infty} 2^{-j\alpha}$. In particular, there exists m so that if $y \in \bigcup_{n=1}^{\infty} V_n$ then

$$d(y,V_m) \leq \delta/4.$$

Pick z in V_n with $\|y-z\| \leq \delta/4$. Now $z = T((1/k)(x_1+\ldots+x_k))$ where $1 \leq k \leq m$. Hence $z = T(u)$ where

$$\|u\|^p \leq k^{1-p} \leq m^{1-p}$$

since $\|\cdot\|^p$ is an F-norm. Hence there exists $x \in X$ with $\|x\| \leq m^{1-p}\delta^{-1}$ and

$$\|y-T(x)\| < 1/2$$

as required.

THEOREM 4.14. If X is a locally bounded F-space such that $\hat{X}$ is a B-convex Banach space, then X is locally convex, i.e., $X = \hat{X}$.

Proof. Consider the natural inclusion map $J : X \to \hat{X}$. Then $\overline{\text{co}}\ J(B_X)$ is the unit ball of $\hat{X}$. By Theorem 4.13 J is an open mapping, i.e., $X = \hat{X}$.

REMARK 1. The spaces ℓ_p for $1 < p < \infty$ are B-convex.

REMARK 2. If $\hat{X} \approx \ell_2$, then $X \approx \ell_2$.

REMARK 3. If $T : \ell_1 \to \ell_2$ is a quotient map, then $T(\ell_p) = \ell_2$ for any p, $0 < p < 1$.

CHAPTER 5
THREE SPACE PROBLEMS

1. Introduction; K-spaces

A three space problem is a question of the following type: Suppose (P) is some property of an F-space, and X is an F-space with a closed subspace N such that both N and X/N have property (P). Does X necessarily have property (P)? In the category of Banach spaces, a typical example is to take the property (P) to be reflexivity. In this case, if N and X/N are reflexive, X is reflexive.

In general, if X has property (P) whenever N and X/N have property (P), we say that the property (P) is a <u>three-space property</u>. In the category of Banach spaces, most general properties that come readily to mind (e.g. injectivity, projectivity, B-convexity, etc.) are three-space properties. However, there are some exceptions, which we will mention later on.

Returning to the category of F-spaces, let us consider the property that X* separates the points of X. (cf. Ribe [1971]). Using a result from Chapter 4 we can easily show that this is <u>not</u> a three-space property. Indeed, for $0 < p < 1$, let M be a weakly closed subspace of ℓ_p which does not have HBEP and let N be the kernel of some continuous linear functional on M which does not extend to ℓ_p. (See Chapter 4.) Then $X = \ell_p/N$ cannot have the point separation property; in fact $M/N = L$ is a subspace of dimension one which does not have HBEP. However L and X/L ($\approx \ell_p/M$) both have the point separation property.

We now state the main problem of this chapter:

PROBLEM 5.1. Is local convexity a three-space property?

More precisely, if X is an F-space with a locally convex subspace M such that X/M is locally convex, must X be locally convex?

This problem was discussed in detail by S. Dierolf [1974], who showed that if one can construct a counterexample, then it can be done with $\dim M = 1$. The problem is thus to find a nonlocally convex F-space with an uncomplemented subspace of dimension one, such that the quotient is locally convex.

From this point it is natural to make the following definition (which was originally motivated by very different considerations):

DEFINITION. An F-space X is a K-<u>space</u> if whenever Y is an F-space and L is a subspace of Y with dimension one such that $Y/L \approx X$, then L is complemented in Y, i.e. $Y \approx L \oplus X$. Problem 5.1 thus becomes: is every locally convex F-space a K-space? As we shall see, the answer to this question is negative. In fact, independently and at approximately the same time, Kalton [1978b], Ribe [1979], and Roberts [1977b] showed that the Banach space ℓ_1 is not a K-space. Later we shall present the Ribe example, which in many ways is the most natural of these examples.

The negative solution to Problem 5.1 also provides, interestingly enough, a negative solution to Problem 4.2, discussed in the previous chapter. Another point of interest, which we shall not pursue here, is that the Ribe space can be modified to provide a counterexample to the three-space problem for Hilbert space. This problem, suggested by Palais, was to decide whether the property of being isomorphic to ℓ_2 is a three-space property. A counterexample was constructed by Enflo, Lindenstrauss and Pisier [1975]. However, a rather easier counterexample can be made using the ideas of Ribe's construction (see Kalton-Peck [1979c]).

In spite of the fact that local convexity is not a

three-space property, the notion of a K-space turns out to be of some general interest. This interest springs from the fact that many classical spaces are K-spaces. In fact we shall show later on (cf. Kalton [1978b]) that the function spaces L_p and the sequence spaces ℓ_p are K-spaces provided $0 < p < \infty$ and $p \neq 1$. The space L_0 is also a K-space (Kalton-Peck [1979b]) but the proof of this is deferred to the next chapter.

We point out at this juncture a relationship between these ideas and results of the previous chapter.

THEOREM 5.1. An F-space X is a K-space if and only if whenever Y is an F-space and N is a closed subspace of Y with $Y/N \approx X$, then N has HBEP.

Proof. One direction is immediate from the definition. For the converse, suppose X is a K-space, $Y/N \approx X$ and $\varphi \in N^*$. Let $M = \ker \varphi$; then N/M has dimension 1 and $(Y/M)/(N/M) = Y/M \approx X$. Since X is a K-space, N/M is complemented in Y/M and it is easy to deduce that φ can be extended to a continuous linear functional on Y. □

REMARK. In view of our promised results on the spaces ℓ_p and L_p for $p \neq 1$, it follows from 4.2 that the kernel M of a quotient mapping from ℓ_p to L_p has HBEP in ℓ_p. Earlier M was exhibited as a PCWD subspace. Thus in fact M has the "unique"-HBEP.

THEOREM 5.2. Suppose X is a K-space and N is a closed subspace of X. Then X/N is a K-space if and only if N has HBEP.

Proof. One direction follows from 5.1. For the converse suppose N has HBEP. Suppose Y is an F-space with a subspace L of dimension one such that $Y/L \approx X/N$. Thus we have the diagram

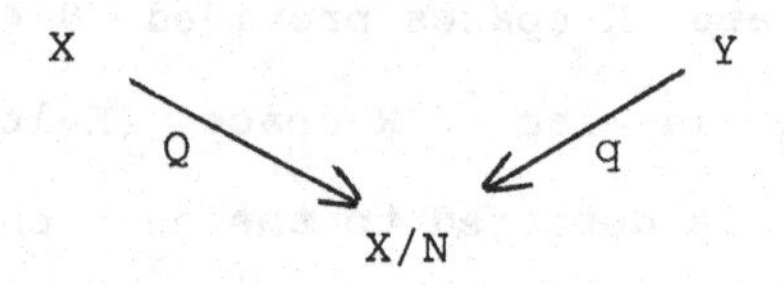

where q and Q are quotient maps, $\ker Q = N$, $\ker q = L$. Now let Z be the subspace of $X \oplus Y$ of all (x,y) such that $Qx = qy$.

We need to show that L has HBEP in Y. Suppose $\varphi \in L^*$. First consider the map $P : Z \to X$ given by

$$P(x,y) = x.$$

P is a surjection and $\ker P = \{0\} \times L$. We can therefore induce a linear functional $\tilde{\varphi}$ on $\ker P$ by

$$\tilde{\varphi}(0,y) = \varphi(y) \qquad\qquad y \in L.$$

However X is a K-space and hence $\ker P$ has the HBEP. Thus there exists $\psi \in Z^*$ with

$$\psi(0,y) = \varphi(y) \qquad\qquad y \in L.$$

Now consider for $x \in N$

$$\tilde{\psi}(x) = \psi(x,0).$$

As N has HBEP, there exists an extension σ of $\tilde{\psi}$ to X. Thus we can define $\rho \in Z^*$ by

$$\rho(x,y) = \psi(x,y) - \sigma(x,0).$$

Then if $P_1 : Z \to Y$ is defined by $P_1(x,y) = y$, P_1 is onto, $\ker P_1 = N \times \{0\}$ and $\rho | \ker P_1 = 0$. Hence ρ induces an element $\tilde{\rho}$ in $Y^* : \tilde{\rho}(y) = \rho(x,y)$, where $(x,y) \in Z$. Now for $y \in L$,

$$\tilde{\rho}(y) = \rho(0,y) = \psi(0,y) = \varphi(y),$$

so $\tilde{\rho}$ extends φ. □

2. <u>Quasi-linear maps and K-spaces</u>

Let X_0 be a (not necessarily complete) quasi-normed space. We now give a construction of a quasi-Banach space Y with the following property: there is a one-dimensional subspace $\mathbf{R}$ of Y and a quotient map q of Y onto the completion of X_0 such that $\ker q = \mathbf{R}$.

DEFINITION. A <u>quasi-linear map</u> from X_0 to $\mathbf{R}$ is a map F satisfying

(a) $F(\lambda x) = \lambda F(x)$, $\lambda \in \mathbf{R}$, $x \in X_0$,

(b) $|F(x_1+x_2)-F(x_1)-F(x_2)| \leq K(\|x_1\|+\|x_2\|)$,

for some constant K and all x_1, x_2 in X_0.

Given a quasi-linear map $F : X_0 \to \mathbf{R}$, we define $\mathbf{R} \oplus_F X_0$ to be the algebraic direct sum $\mathbf{R} \oplus X_0$ with quasi-norm

$$\|(\lambda,x)\| = |\lambda-F(x)| + \|x\|.$$

We first check that the above actually defines a quasi-norm. Indeed, the only point to check is that

$$\|(\lambda_1,x_1)+(\lambda_2,x_2)\| \leq C(\|(\lambda_1,x_1)\|+\|(\lambda_2,x_2)\|)$$

for some constant C and all $\lambda_1,\lambda_2 \in \mathbb{R}$, $x_1,x_2 \in X_0$. Now

$$\|x_1+x_2\| \leq C_1(\|x_1\|+\|x_2\|)$$

for some constant C_1, and all x_1,x_2 in X, and hence

$$\begin{aligned}\|(\lambda_1,x_1)+(\lambda_2,x_2)\| &\leq |\lambda_1+\lambda_2-F(x_1+x_2)| + C_1(\|x_1\|+\|x_2\|)\\ &\leq |\lambda_1-F(x_1) + \lambda_2-F(x_2)|\\ &\quad + |F(x_1+x_2)-F(x_1)-F(x_2)|\\ &\quad + C_1(\|x_1\|+\|x_2\|)\\ &\leq C_1(\|(\lambda_1,x_1)\|+\|(\lambda_2,x_2)\|)\\ &\quad + K(\|x_1\|+\|x_2\|)\\ &\leq (C_1+K)(\|(\lambda_1,x_1)\|+\|(\lambda_2,x_2)\|)\\ &\quad\quad \text{(since } C_1 \geq 1).\end{aligned}$$

The map $(\lambda,x) \to x$ is an open surjection of $\mathbb{R} \oplus_F X_0$ onto X_0, since

(i) $\|x\| \leq \|(\lambda,x)\|$ for all λ, and
(ii) $\|x\| = \|(F(x),x)\|$.

Now suppose X is a quasi-Banach space and X_0 is a fixed dense subspace of X. Suppose that F is a quasi-linear map from X_0 to $\mathbb{R}$. We define $\mathbb{R} \oplus_F X$ to be the completion of $\mathbb{R} \oplus_F X_0$. Let q be the map of $\mathbb{R} \oplus_F X$ into X which is the continuous extension of the map $q(\lambda,x) = x$, $\lambda \in \mathbb{R}$, $x \in X_0$.

LEMMA 5.3. (a) The map q is an open surjection of $\mathbb{R} \oplus_F X$ onto X;

(b) $\ker q$ equals $\{(\lambda,0) : \lambda \in \mathbb{R}\}$.

<u>Proof</u>. (a) This is almost immediate: if $x \in X$, let (x_n) be a sequence in X_0 converging to x. Since $q|_{R \oplus_F X_0}$ is an open map onto X_0, it is not hard to see that there is a Cauchy sequence (z_n) in $R \oplus_F X_0$ with $q(z_n) = x_n$. Then x is equal to $q(z)$, where $z = \lim_n z_n$. For (b), the containment one way is clear. Now assume (λ_n, x_n) is a sequence in $R \oplus_F X_0$ converging to z with $q(z) = 0$. Then $\|x_n\| \to 0$, so $(F(x_n), x_n) \to 0$ in $R \oplus_F X_0$, and now $(\lambda_n - F(x_n), 0) \to z$. Hence z is in $R \oplus \{0\}$. □

Thus we have described a means of constructing a quasi-Banach space Y whose quotient by a line is isomorphic to X. We now show that this is essentially the only way.

LEMMA 5.4. Suppose X and Y are quasi-Banach spaces and that $Q : Y \to X$ is an open surjection with $\dim(\ker Q) = 1$. Suppose X_0 is some given dense subspace of X. Then there are a quasi-linear map $F : X_0 \to R$ and an isomorphism J from $R \oplus_F X$ onto Y so that the following diagram commutes:

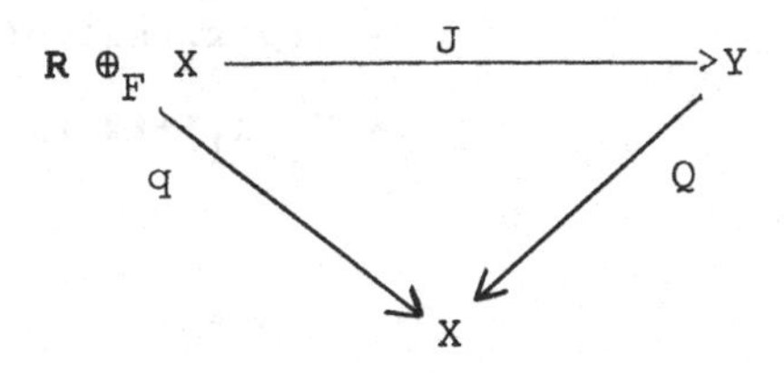

<u>Proof</u>. Let $\ker Q = \{\lambda u : \lambda \in \mathbb{R}\}$, where $\|u\| = 1$. By the open mapping theorem there is a constant C such that if $\|x\| \leq 1$ there is $y \in Y$ with $\|y\| < C$ and $Qy = x$. We can then define a (possibly discontinuous) map $\rho : X_0 \to Y$ so that

$$\rho(\lambda x) = \lambda\rho(x), \qquad x \in X_0, \qquad \lambda \in \mathbb{R},$$
$$Q\rho(x) = x, \qquad x \in X_0.$$
$$\|\rho(x)\| \leq C\|x\|, \qquad x \in X_0.$$

Now let $\varphi : X_0 \to Y$ be any <u>linear</u> map so that

$$Q\varphi(x) = x, \qquad x \in X_0.$$

(This can be done very simply by using a Hamel basis for X_0.)

Now $\rho(x) - \varphi(x) \in \ker Q$, so we can define $F : X_0 \to \mathbb{R}$ by

$$F(x)u = \rho(x) - \varphi(x).$$

Then F is quasi-linear since

$$\begin{aligned}|F(x_1+x_2)-F(x_1)-F(x_2)| &= \|(F(x_1+x_2)-F(x_1)-F(x_2))u\| \\ &= \|(\rho(x_1+x_2)-\rho(x_1)-\rho(x_2))u\| \\ &\leq K(\|x_1\|+\|x_2\|)\end{aligned}$$

for some constant K and all $x_1, x_2 \in X_0$.

Define $J_0 : \mathbb{R} \oplus_F X_0 \to Y$ by

$$J_0(\lambda, x) = \lambda u + \varphi(x).$$

Then if $(\lambda_n, x_n) \to 0$, we have $\|x_n\| \to 0$ so $(F(x_n), x_n) \to 0$ and then $\lambda_n - F(x_n) \to 0$. Hence

$$\lambda_n u - \rho(x_n) + \varphi(x_n) \to 0$$

and

$$\rho(x_n) \to 0,$$

so that $J_0(\lambda_n, x_n) \to 0$, i.e. J_0 is continuous. Conversely if $J_0(\lambda_n, x_n) \to 0$, then

$$\|\lambda_n u + \varphi(x_n)\| \to 0$$

and $Q(\lambda_n u + \varphi(x_n)) = x_n \to 0$. Hence $\rho(x_n) \to 0$ and so

$$\lambda_n u + \varphi(x_n) - \rho(x_n) \to 0$$

i.e.

$$\lambda_n u - F(x_n)u \to 0.$$

As $(F(x_n), x_n) \to 0$ this implies $(\lambda_n, x_n) \to 0$, i.e. J_0 is an isomorphism, and extends to an isomorphism between $R \oplus_F X$ and Y. For $\lambda \in R$, $x \in X_0$, $QJ(\lambda, x) = Q(\lambda u + \varphi(x)) = x = q(\lambda, x)$. Hence $QJ = q$ when we pass to the completions. □

PROPOSITION 5.5 In order that the kernel of the natural surjection of $R \oplus_F X$ onto X be complemented, it is necessary and sufficient that there is a linear map $G : X_0 \to R$ with $\sup_{\|x\| \leq 1} |F(x) - G(x)| = L < \infty$.

Proof. If such a G exists, define $P : R \oplus_F X_0 \to R \oplus_F X_0$ by

$$P(\lambda,x) = (\lambda-G(x),0).$$

It is immediate that $P^2 = P$; P is continuous since

$$|\lambda-G(x)| \leq |\lambda-F(x)| + |F(x)-G(x)|$$
$$\leq (L+1)\|(\lambda,x)\|.$$

Then P extends to a projection of $\mathbf{R} \oplus_F X$ onto $\mathbf{R} \oplus \{0\}$. Conversely if such a P exists its restriction to $\mathbf{R} \oplus_F X_0$ must take the above form, with G a linear map from X_0 to $\mathbf{R}$. Essentially the argument above shows that for some constant A,

$$|F(x)-G(x)| \leq A\|x\|, \qquad x \in X_0.$$

(Note that G need <u>not</u> be continuous.) □

COROLLARY 5.6. Let X be a quasi-Banach space and suppose X_0 is a dense subspace of X. In order that X be a K-space it is necessary and sufficient that whenever $F : X_0 \to \mathbf{R}$ is quasi-linear there is a linear map $G : X_0 \to \mathbf{R}$ with

$$\sup_{\|x\|\leq 1} |F(x)-G(x)| < \infty.$$

<u>Proof</u>. This is immediate from Proposition 5.5 and Lemma 5.4. □

3. <u>The spaces</u> ℓ_p <u>and</u> L_p <u>for</u> $0 < p < 1$

THEOREM 5.7. The spaces ℓ_p for $0 < p < 1$ are K-spaces.

<u>Proof</u>. We shall require a finite-dimensional lemma.

LEMMA 5.8. Let G be a quasi-linear map from ℓ_p^m into $\mathbb{R}$. Then

$$|G(\sum_{i=1}^m t_i e_i) - \sum_{i=1}^m t_i G(e_i)| \leq \Delta(G) \sum_{i=1}^m (2/i)^{1/p} \| \sum_{i=1}^m t_i e_i \|,$$

where $\Delta(G)$ is the least constant K such that

$$|G(x+y)-G(x)-G(y)| \leq K(\|x\|+\|y\|) \qquad x,y \in \ell_p^m.$$

<u>Proof</u> (of the Lemma). The proof is by induction. The lemma is trivial if $m = 1$. Let us now assume it is true for $1 \leq m < n$. Choose $x = \sum t_i e_i \in \ell_p^n$, and select j,k, with $j \neq k$, so that

$$|t_j|^p + |t_k|^p \leq 2\|x\|^p/n.$$

Let $u = t_j e_j + t_k e_k$. If $u = 0$ then x belongs to the linear span of $(n-2)$ basis vectors and the lemma easily follows from the inductive hypothesis. Suppose $u \neq 0$, and let X be the linear span of $\{e_i : i \neq j,k\} \cup \{\|u\|^{-1}u\}$. By the inductive hypothesis,

$$|G(x) - (\sum_{i\neq j,k} t_i G(e_i)) - G(u)| \leq \Delta(G) \sum_{i=1}^{n-1} (2/i)^{1/p} \|x\|$$

and

$$\begin{aligned}|G(u)-t_jG(e_j)-t_kG(e_k)| &\le \Delta(G)(|t_j|+|t_k|)\\ &\le \Delta(G)(|t_j|^p+|t_k|^p)^{1/p}\\ &\le \Delta(G)(2/n)^{1/p}\|x\|.\end{aligned}$$

Hence

$$|G(x) - \sum_{i=1}^{n} t_iG(e_i)| \le \Delta(G) \sum_{i=1}^{n}(2/i)^{1/p}\|x\|$$

and this concludes the proof of the lemma. □

We now complete the proof of the theorem. Let X_0 be the space of finitely non-zero elements in ℓ_p. Suppose $F : X_0 \to R$ is a quasi-linear map. Since

$$\sum_{i=1}^{\infty} (2/i)^{1/p} = M < \infty$$

we conclude that for $x \in X_0$,

$$|F(x) - \sum_{i=1}^{\infty} x_iF(e_i)| \le \Delta(F)M\|x\|.$$

Thus $G(x) = \sum_{i=1}^{\infty} x_iF(e_i)$ is a linear map from $X_0 \to R$ which satisfies the condition of 5.6. □

Theorem 5.2 and the above now imply

COROLLARY 5.9. Suppose X is a locally bounded F-space and $q : \ell_p \to X$ is any surjection. Then X is a K-space if and only if $\ker q$ has HBEP.

If X is separable and p-convex, such a surjection always exists by Proposition 2.10.

COROLLARY 5.10. The spaces L_p for $0 < p < 1$ are K-spaces.

Proof. Let $q : \ell_p \to L_p$ be any quotient map. Let $E_n \subset L_p$ be the span of the characteristic functions $\chi_{k,n} = 1_{[k\cdot 2^{-n},(k+1)\cdot 2^{-n})}$ for $0 \leq k < 2^n$.

By the open mapping theorem there exists $M < \infty$ such that if $f \in L_p$ there is $h \in \ell_p$ with $q(h) = f$ and $\|h\| \leq M\|f\|$. For each k,n, choose $h_{k,n} \in \ell_p$ with $qh_{k,n} = \chi_{k,n}$ and

$$\begin{aligned}\|h_{k,n}\| &\leq M\|\chi_{k,n}\| \\ &= M\cdot 2^{-n/p}.\end{aligned}$$

Define $S_n : E_n \to \ell_p$ by

$$S_n\chi_{k,n} = h_{k,n}.$$

Then $\|S_n\| \leq M$.

Now suppose $\varphi \in (\ker q)^*$. Then we may define $\varphi_n \in q^{-1}(E_n)^*$ by

$$\varphi_n(x) = \varphi(x-S_nqx)$$

since $q(x-S_nqx) = qx - (qS_n)qx = 0$. Then

$$|\varphi_n(x)| \leq \|\varphi\| \cdot \|x - S_n qx\|$$
$$\leq \|\varphi\|(\|x\|^p + M^p\|q\|^p\|x\|^p)^{1/p}$$
$$= C\|\varphi\| \cdot \|x\|$$

where $C = (1+M^p\|q\|^p)^{1/p}$.

Now let $F \subset \ell_p$ be defined by $F = \bigcup_{n=1}^{\infty} q^{-1}(E_n)$. By an elementary compactness argument there exists a linear map $\psi : F \to R$ with $\psi(x)$ a cluster point of $\{\varphi_n(x)\}_{n>m}$ whenever $x \in q^{-1}(E_m)$. Thus $\psi \in F^*$ and

$$\|\psi\| \leq C\|\varphi\|.$$

As F is dense in ℓ_p, ψ extends to an element of $(\ell_p)^*$, and by construction ψ extends φ. Thus (ker q) has HBEP, so L_p is a K-space. □

COROLLARY 5.11. If M is a closed subspace of L_p and $L_p/M \approx L_p$ then $M^* = \{0\}$.

Proof. L_p/M is a K-space so that M has HBEP i.e. $M^* = \{0\}$. □

In particular, if $0 < \dim M < \infty$, $L_p/M \not\approx L_p$, in striking contrast to the situation when $p > 1$.

4. The Ribe space

Our next objective is to show that the Banach space ℓ_1 is not a K-space, which will in turn establish that local convexity is not a three-space property. The space ℓ_1 is a quotient of ℓ_p for $0 < p < 1$; the above then implies that the kernel of the quotient map fails HBEP.

Since every separable Banach space is a quotient of ℓ_1 by a subspace with HBEP, the existence of any (separable) Banach non-K-space implies that ℓ_1 is not a K-space. (In fact, a compactness argument can be used to eliminate separability in this statement.)

As mentioned earlier, Kalton [1978b], Ribe [1979], and Roberts [1977b] proved independently that ℓ_1 is not a K-space. There is an important tie-in between Roberts's approach and his construction (see Chapter 9) of a compact convex set with no extreme points: in both constructions the central idea is that of "needle point"; appropriate finite-dimensional constructions involving needle points then lead to the results.

We present here the Ribe construction.

LEMMA 5.12. For $s,t \in R$,

$$|s\log|s| + t\log|t| - (s+t)\log|s+t|| \leq |s| + |t|,$$

where by convention $0 \log 0 = 0$.

Proof. First consider the case $0 < s,t$. Then

$$|(s+t)\log|s+t| - s\log|s| - t\log|t||$$
$$= -s \log(s/s+t) - t \log(t/s+t)$$
$$= -(s+t)[(s/s+t)\log(s/s+t)+(t/s+t)\log(t/s+t)]$$
$$\leq (2/e)(s+t)$$
$$\leq s+t.$$

Here we have used that if $0 \leq x \leq 1$, $|x\log x| \leq 1/e$.

Clearly the lemma now follows if s and t have the same sign. For the case of opposite signs, we may suppose $s+t > 0$, $t > 0$, and $s < 0$. Apply the above with s replaced by $-s$, t replaced by $s+t$. Then

$$|s\log|s| + t\log|t| - (s+t)\log|s+t||$$
$$\leq |s| + |s+t|$$
$$\leq |s| + |t|. \quad \square$$

THEOREM 5.13 ℓ_1 is not a K-space.

<u>Proof</u>. Let X_0 be the space of finitely non-zero elements of ℓ_1 and define $F : X_0 \to R$ by

$$F(x) = \sum_{n=1}^{\infty} x_n \log|x_n| - (\sum_{n=1}^{\infty} x_n)\log|\sum_{n=1}^{\infty} x_n|.$$

We shall show that F is quasi-linear and that $R \oplus \{0\}$ is uncomplemented in $R \oplus_F \ell_1$.

A straightforward calculation shows that F is homogeneous. To finish the proof that F is quasi-linear, let x and y be in X_0. By the lemma, for each i we have

$$|(x_i+y_i)\log|x_i+y_i|-x_i\log|x_i|-y_i\log|y_i||$$
$$\leq |x_i| + |y_i|.$$

Summing over i, we obtain

(i) $$|\sum_i (x_i+y_i)\log|x_i+y_i| - \sum x_i\log|x_i| - \sum y_i\log|y_i||$$
$$\leq \|x\| + \|y\|.$$

Again by the Lemma,

(ii) $$|(\sum x_i+y_i)\log|\sum x_i+y_i|)$$
$$- (\sum x_i)\log|\sum x_i| - (\sum y_i)\log|\sum y_i||$$
$$\leq |\sum x_i| + |\sum y_i|$$
$$\leq \|x\| + \|y\|.$$

Inequalities (i) and (ii) now show that

$$|F(x+y)-F(x)-F(y)| \leq 2(\|x\| + \|y\|).$$

Now suppose $R \oplus \{0\}$ is complemented in $R \oplus_F \ell_1$. Then there are a constant M and a linear map $G : X_0 \to R$ with

$$|F(x)-G(x)| \leq M\|x\|, \qquad x \in X_0.$$

We have $F(e_i) = 0$ for all i and hence

$$|G(e_i)| \leq M \text{ for all } i.$$

Hence $|G(x)| \leq M\|x\|$, $x \in X_0$, and so

$$|F(x)| \leq 2M\|x\|, \qquad x \in X_0.$$

However

$$F((1/n)(e_1+\dots+e_n)) = -(1/n)\, n \log n$$
$$= -\log n,$$

a contradiction. This finishes the proof that ℓ_1 is not a K-space. □

REMARKS. The appearance of $\log n$ is not a coincidence, and indeed an examination of the proof of Theorem 5.8 reveals that $\log n$ is the maximal rate of increase possible here.

The space constructed will be referred to as the Ribe space. We next show (following Roberts [1977b] that it cannot have a PCWD subspace although it does fail to be locally convex.

THEOREM 5.14. The Ribe space has no PCWD subspace.

Proof. Suppose X is the Ribe space and $q : X \to \ell_1$ is the natural surjection. Suppose $M \subset X$ is closed and weakly dense. Then $q(M)$ is weakly dense in ℓ_1 and hence also dense in ℓ_1. Now $q^{-1}q(M) = M + \ker q$ is dense in X, but is also closed (since $\dim \ker q = 1$). Hence $M + \ker q = X$, and as $\ker q$ is uncomplemented in X, $M = X$. □

5. _The space_ ℓ_p _for_ $1 < p < \infty$

For a quasi-Banach space $(X, \|\cdot\|)$ we now introduce two sequences $a_n = a_n(X)$ and $b_n = b_n(X)$.

DEFINITION. (a) $a_n = \sup(\|x_1+\dots+x_n\| : \|x_i\| \leq 1)$.

(b) $b_n = \sup\limits_{\|x_i\|\leq 1} \inf\limits_{\theta_i=\pm 1} \|\sum\limits_{i=1}^{n} \theta_i x_i\|$.

Observe that, in general, $0 \leq b_n \leq a_n$, and $n \leq a_n$. If $\|\cdot\|$ is p-subadditive than $a_n \leq n^{1/p}$; if $\|\cdot\|$ is equivalent to a p-convex quasi-norm then $a_n \leq Cn^{1/p}$ for some constant C. Some more properties of (a_n) and (b_n) are given in the next lemma and proposition.

LEMMA 5.15. The sequences (a_n) and (b_n) are monotone increasing and satisfy $a_{mn} \leq a_m a_n$, $b_{mn} \leq b_m b_n$ for $m,n \in \mathbf{N}$.

<u>Proof</u>. The only point we shall check is that $b_{mn} \leq b_m b_n$. Suppose $\{x_{ij} : 1 \leq i \leq m,\ 1 \leq j \leq n\}$ is a set of mn vectors in the unit ball of X. Then for suitable signs $\eta_{ij} = \pm 1$,

$$\left\| \sum_{j=1}^{n} \eta_{ij} x_{ij} \right\| \leq b_n, \qquad i = 1,2,\ldots,m$$

and for suitable $\delta_i = \pm 1$,

$$\left\| \sum_{i=1}^{m} \sum_{j=1}^{n} \delta_i \eta_{ij} x_{ij} \right\| \leq b_m b_n$$

i.e.

$$b_{mn} \leq b_m b_n. \qquad \square$$

PROPOSITION 5.16. (a) If $a_n = O(n)$, X is locally convex,

(b) If $b_n = o(n)$, X is locally convex.

<u>Proof</u>. (a) Suppose $a_n \leq Cn$, and $\|x_i\| \leq 1$, $i = 1,2,\ldots,m$,

and $\alpha_i > 0$, $\sum_{i=1}^{m} \alpha_i = 1$. Choose $\beta_i^{(n)} > 0$ rational so that $\beta_i^{(n)} \to \alpha_i$, $\sum \beta_i^{(n)} = 1$. Then for each n, there exists a common demoninator N_n so that

$$\beta_i^{(n)} = M_{n,i}/N_n$$

where $M_{n,i}, N_n \in \mathbf{N}$. Thus

$$\begin{aligned} \| \sum_{i=1}^{m} \beta_i^{(n)} x_i \| &= 1/N_n \| \sum_{i=1}^{m} M_{n,i} x_i \| \\ &\leq (1/N_n)\, a_{N_n} \\ &\leq C. \end{aligned}$$

Hence $\sum \alpha_i x_i$ belongs to the closure of the set $\{x: \|x\| \leq C\}$, and so the convex hull of the unit ball of X is bounded, i.e., X is locally convex. □

(b) We show $a_n = O(n)$. To do this, it will suffice to consider the case when the quasi-norm is r-subadditive for some r, $0 < r < 1$. Let $(x_i : 1 \leq i \leq 2n)$ be any $2n$ vectors in the unit ball of X. Then for suitable signs $\theta_i = \pm 1$ we have

$$\|\theta_1 x_1 + \ldots + \theta_{2n} x_{2n}\| \leq b_{2n}$$

and we may suppose that $\operatorname{card}\{i : \theta_i = +1\} \geq n$. Then

$$\sum_{i=1}^{2n} x_i = \sum_{i=1}^{2n} \theta_i x_i + 2 \sum_{\theta_i = -1} x_i$$

so that

$$\|\sum_{i=1}^{2n} x_i\|^r \leq b_{2n}{}^r + 2^r a_n{}^r$$

i.e.,

$$a_{2n}{}^r \leq b_{2n}{}^r + 2^r a_n{}^r.$$

Let $A_n = a_n/n$ and $B_n = b_n/n$; then

(i) $$A_{2n}{}^r \leq B_{2n}{}^r + A_n{}^r.$$

It will suffice to show that (A_{2^n}) is bounded, since it then follows from the monotonicity of (a_n) that (A_n) is bounded. To do this, it suffices to show that $\sum B_{2^n}{}^r < \infty$, by (i) above. (Note that $a_n \leq a_{2n}/2$, so $A_n \leq A_{2n}$.)

Now fix k sufficiently large; then

$$b_{2^k} < 2^k$$

i.e.

$$B_{2^k} = \beta^k$$

where $\beta < 1$. Hence, by Lemma 5.15,

$$b_{2^{k\ell}} \leq \beta^{k\ell} 2^{k\ell} \qquad \ell = 1,2,\ldots$$

and in general

$$B_{2^n} \leq C\beta^n \quad \text{for} \quad n \in \mathbb{N}.$$

Hence $\sum B_{2n}{}^r < \infty$ as required. □

Thus a quasi-Banach space for which $b_n = o(n)$ is automatically a Banach space. Banach spaces with this property are called B-convex (see Chapter 4, section 4) and have been widely studied in connection with probability in Banach spaces. Several equivalent formulations of B-convexity are known. For our purposes, the most useful notion is that of type.

DEFINITION. A quasi-Banach space X is of type p where $0 < p \leq 2$ if

$$E(\|\sum_{i=1}^{n} \epsilon_i x_i\|^p) = \int_\Omega \|\sum \epsilon_i(\omega) x_i\|^p dP(\omega) \leq C^p(\sum_{i=1}^{n} \|x_i\|^p)$$

for some constant C and all $x_1, \ldots, x_n$ in X. Here, the $\epsilon_i = \epsilon_i(\omega)$ are independent Bernoulli random variables on a probability space (Ω, P) with $P(\epsilon_i = +1) = P(\epsilon_i = -1) = 1/2$. (One may choose $\Omega = [0,1]$ and $\epsilon_i = r_i$, the i'th Rademacher function on $[0,1]$.)

It turns out that a Banach space is B-convex if and only if it is type p for some $p > 1$ (Pisier [1974]). Khintchine's inequality says that the real line is of type 2 and of type p for no $p > 2$; hence for $p > 2$ the notion of type p is uninteresting. The fact that a Hilbert space is of type 2 is a direct consequence of the parallelogram law; the constant C occurring in the definition of type is 1 in this case. Important examples of spaces which are not of type p for any $p > 1$ (i.e., are not B-convex) are ℓ_1 and ℓ_∞.

We illustrate the notion by computing the type of ℓ_p and L_p.

THEOREM 5.17. The spaces ℓ_p and L_p are of type q where $q = \min(2,p)$ for $1 \leq p < \infty$.

<u>Proof</u>. It suffices to consider $L_p = L_p(0,1)$. First consider the case $1 \leq p \leq 2$. Then

$$\int_\Omega \|\sum_{i=1}^n \epsilon_i(\omega)x_i\|^p dP(\omega)$$
$$= \int_\Omega \int_0^1 |\sum_{i=1}^n \epsilon_i(\omega)x_i(t)|^p dt\, dP(\omega)$$
$$= \int_0^1 \int_\Omega |\sum_{i=1}^n \epsilon_i(\omega)x_i(t)|^p dP(\omega)dt$$
$$\leq \int_0^1 (\int_\Omega |\sum_{i=1}^n \epsilon_i(\omega)x_i(t)|^2 dP(\omega))^{p/2} dt$$
$$= \int_0^1 (\Sigma|x_i(t)|^2)^{p/2} dt$$
$$\leq \int_0^1 \Sigma|x_i(t)|^p dt$$
$$= \Sigma \|x_i\|^p.$$

Next consider the case $2 < p < \infty$. We shall need the Khintchine inequality which states that there is A_p such that

$$\int_\Omega |\sum_{i=1}^n \eta_i\epsilon_i(\omega)|^p dP(\omega) \leq A_p^p(\Sigma|\eta_i|^2)^{p/2}$$

for any $\eta_1,\dots,\eta_n \in \mathbb{R}$. Now

$$\int_\Omega \|\sum_{i=1}^n \epsilon_i(\omega)x_i\|^2 dP(\omega) = \int_\Omega (\int_0^1 |\Sigma\epsilon_i(\omega)x_i(t)|^p dt)^{2/p} dP(\omega)$$
$$\leq (\int_\Omega \int_0^1 |\Sigma\epsilon_i(\omega)x_i(t)|^p dt\, dP(\omega))^{2/p}$$

(since $2/p < 1$)

$$= (\int_0^1 \int_\Omega |\Sigma\epsilon_i(\omega)x_i(t)|^p dP(\omega)dt)^{2/p}$$
$$\leq (\int_0^1 A_p^p(\sum_{i=1}^n |x_i(t)|^2)^{p/2} dt)^{2/p}$$

$$\leq A_p^{\,2} \sum_{i=1}^{n} \left(\int_0^1 |x_i(t)|^p dt \right)^{2/p}$$

(by the subadditivity of the $L_{p/2}$-norm)

$$= A_p^{\,2} \sum_{i=1}^{n} \|x_i\|^2. \quad \square$$

THEOREM 5.18. Let X be a Banach space of type $p > 1$ (i.e., X is B-convex); then X is a K-space.

In particular ℓ_p and L_p are K-spaces for $1 < p < \infty$.

<u>Proof</u>. Let X_0 be a dense subspace of X and let $F : X_0 \to \mathbb{R}$ be any quasi-linear map. We shall show that $Y_0 = \mathbb{R} \oplus_F X_0$ is necessarily locally convex, which will imply the result.

For $x_1, \ldots, x_n$ in X_0 define

$$\Delta(x_1, \ldots, x_n) = F\Big(\sum_{i=1}^{n} x_i\Big) - \sum_{i=1}^{n} F(x_i).$$

Observe the following property of Δ: if $(A_j : 1 \leq j \leq m)$ is a partition of $\{1, 2, \ldots, n\}$ and

$$u_j = \sum_{i \in A_j} x_i,$$

then

$$\Delta(x_1, \ldots, x_n) = \Delta(u_1, \ldots, u_m) + \sum_{j=1}^{m} \Delta(x_i : i \in A_j).$$

Now X is of type $p > 1$, so

$$E(\|\sum_{i=1}^{n}\epsilon_i x_i\|^p)^{1/p} \le C(\sum_{i=1}^{n}\|x_i\|^p)^{1/p}, \qquad x_1,\ldots,x_n \in X.$$

Let us define δ_n to be the least constant so that

$$(E|\Delta(\epsilon_1 x_1,\ldots,\epsilon_n x_n|^p)^{1/p} \le \delta_n(\sum_{i=1}^{n}\|x_i\|^p)^{1/p}$$
$$x_1,\ldots,x_n \quad \text{in} \quad X.$$

Let $(x_{ij} : 1 \le i \le m,\ 1 \le j \le n)$ be a collection of mn elements of X and suppose $(\epsilon_{ij} = \epsilon_{ij}(\omega): 1 \le i \le m,$ $1 \le j \le n)$ are independent Bernoulli random variables on some probability space (Ω,P). Then, by symmetry,

$$E(|\sum_{j=1}^{n}\Delta(\epsilon_{ij}x_{ij} : 1 \le i \le m)|^p)$$
$$= E(|\sum_{j=1}^{n}\eta_j\Delta(\epsilon_{ij}x_{ij} : 1 \le i \le m)|^p)$$

where $(\eta_j = \eta_j(t) : 1 \le j \le n)$ is a further sequence of Bernoulli random variables, mutually independent and independent of (ϵ_{ij}).

Now by Fubini's theorem, and since R has type p with constant one,

$$E(|\sum_{j=1}^{n}\Delta(\epsilon_{ij}x_{ij} : 1 \le i \le m)|^p)$$
$$\le \sum_{j=1}^{n}E(|\Delta(\epsilon_{ij}x_{ij} : 1 \le i \le m)|^p)$$
$$\le \delta_m^p\sum_{j=1}^{n}\sum_{i=1}^{m}\|x_{ij}\|^p.$$

For $\omega \in \Omega$, let $u_j(\omega) = \sum_{i=1}^{n} \epsilon_{ij}(\omega)x_{ij}$. Again, arguing by symmetry,

$$E(|\Delta(u_1,u_2,\ldots,u_n)|^p = E(|\Delta(\eta_1 u_1,\ldots,\eta_n u_n)|^p)$$

$$\leq \delta_n{}^p E(\sum_{j=1}^{n} \|u_j\|^p)$$

$$\leq C^p \delta_n{}^p \sum_{j=1}^{n} \sum_{i=1}^{m} \|x_{ij}\|^p.$$

Now

$$(E(|\Delta(\epsilon_{ij}x_{ij} : 1 \leq i \leq m, 1 \leq j \leq n)|^p))^{1/p}$$

$$\leq (E(|\Delta(u_1,\ldots,u_n)|^p))^{1/p}$$

$$+ (E(|\sum_{j=1}^{n} \Delta(\epsilon_{ij}x_{ij} : 1 \leq i \leq m)|^p))^{1/p}$$

$$\leq (C\delta_n + \delta_m)(\sum_{i=1}^{m} \sum_{j=1}^{n} \|x_{ij}\|^p)^{1/p}$$

It follows that

$$\delta_{mn} \leq C\delta_n + \delta_m;$$

since (δ_n) is monotone increasing, this implies that $\delta_n \leq C_1 \log n$ for some constant C_1.

Now suppose (y_i) $1 \leq i \leq n$, are elements of $Y_0 = R \oplus_F X_0$, with $\|y_i\| \leq 1$ for all i. Let

$$y_i = (\alpha_i, x_i), \quad \alpha_i \in \mathbb{R}, \quad x_i \in X_0,$$

with

$$|\alpha_i - F(x_i)| + \|x_i\| \leq 1.$$

We now have

$$\Big(E\|\sum_{i=1}^{n} \epsilon_i y_i\|^p\Big)^{1/p}$$

$$= \Big(E\Big(|\sum_{i=1}^{n} \epsilon_i \alpha_i - F(\sum_{i=1}^{n} \epsilon_i x_i|) + \|\sum_{i=1}^{n} \epsilon_i x_i\|\Big)^p\Big)^{1/p}$$

$$\leq \Big(E\Big(|\sum_{i=1}^{n} \epsilon_i(\alpha_i - F(x_i))|^p\Big)\Big)^{1/p}$$

$$+ \Big(E(|\Delta(\epsilon_1 x_1, \ldots, \epsilon_n x_n)|)^p\Big)^{1/p}$$

$$+ \Big(E\Big(\|\sum_{i=1}^{n} \epsilon_i x_i\|^p\Big)\Big)^{1/p}$$

$$\leq (C+1)n^{1/p} + C_1(\log n)n^{1/p}.$$

Hence $b_n = b_n(Y_0)$ satisfies

$$b_n/n \leq (C + 1 + C_1 \log n)n^{1/p - 1},$$

and by Proposition 5.16(b), Y_0 is locally convex. Hence Y, the completion of Y_0, is locally convex. □

REMARK. Since $Y \approx \mathbb{R} \oplus X$, Y is also of type p, i.e. (δ_n) is in fact bounded.

REMARKS. (1) For $p < 1$, a quasi-Banach space is of type p if and only if $\|\cdot\|$ is equivalent to a p-convex quasi-norm (Kalton [1981a]).

(2) For more details on type, Rademacher averaging, etc., see for example Garling [1977].

PROBLEM 5.2. Determine precise conditions on a Banach space so that it is a K-space.

Recently it was shown that c_0 is a K-space (Kalton-Roberts, to appear). A reasonable conjecture appears to be that X is a K-space if and only if X does not contain uniformly complemented ℓ_1^n's.

CHAPTER 6
LIFTING THEOREMS

1. Introduction

In the preceding two chapters we have seen that the notion of a K-space arises naturally from problems concerning the Hahn-Banach Theorem. However K-spaces were originally introduced in Kalton-Peck (1979b) for quite different reasons. We show first that they satisfy a lifting property for operators.

THEOREM 6.1. Suppose X is an F-space. Then X is a K-space if and only if it has the property:

(P) whenever Y is an F-space and $E \subset Y$ is a finite-dimensional subspace of Y then any operator $T : X \to Y/E$ can be lifted to an operator $T_1 : X \to Y$ (so that $QT_1 = T$ where $Q : Y \to Y/E$ is the quotient map).

Proof. Suppose X has property (P) and Z is a twisted sum of $\mathbb{R}$ and X. Then the identity map $i : X \to X$ may be lifted to a map $j : X \to Z$ and it immediately follows that $Z \cong X \oplus \mathbb{R}$.

Conversely suppose X is a K-space and that $T : X \to Y/E$ is an operator, where Y is an F-space and $\dim E < \infty$. We introduce the space $G \subset X \oplus Y$ of all (x,y) so that $Tx = Qy$. We thus have the diagram

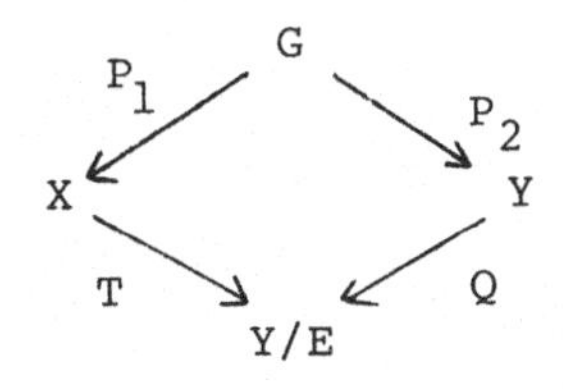

where $P_1(x,y) = x$ and $P_2(x,y) = y$. Now $\ker P_1 = \{(0,y) : y \in E\}$ and so $\dim \ker P_1 < \infty$. However P_1 is a surjection and so $\ker P_1$ has HBEP, and is thus complemented in G. This means there exists $S : X \to G$ so that $P_1S(x) = x$ for $x \in X$. Now P_2S lifts T since $QP_2Sx = Tx$. □

In this chapter we shall establish first a lifting theorem for $L_0(0,1)$ which will imply that L_0 is a K-space. We then revisit the spaces $L_p(0,1)$ for $0 < p < 1$ and establish some similar lifting theorems for them. Most of these results are in Kalton-Peck [1979b].

2. Lifting Theorems for L_0

Let X be any F-space. For $x \in X$ we define $\sigma : X \to [0,\infty]$ by:

$$\sigma(x) = \sup_{n \in \mathbf{N}} \|nx\|.$$

Thus $\sigma(x) = \infty$ is possible. If $X = L_0(0,1)$ with its usual F-norm

$$\|f\| = \int_0^1 |f(t)|/(1+|f(t)|)dt,$$

then

$$\sigma(f) = \lambda(\text{supp } f)$$

where $\operatorname{supp} f = \{t : |f(t)| > 0\}$.

If Y is a subspace of X we shall define $\sigma(Y) = \sup(\sigma(y) : y \in Y)$.

Before proving our main results we need a couple of elementary lemmas.

LEMMA 6.2 Let X be an F-space and let B be a closed locally bounded subspace of X. Let $Q : X \to X/B$ denote the quotient map. Suppose $\delta > 0$ is chosen so that the set $\{b \in B : \|b\| \leq \delta\}$ is bounded.

Then, if $\xi \in X/B$ and $\sigma(\xi) \leq \delta/3$, there is a unique $x \in X$ with $Qx = \xi$ and $\sigma(x) \leq \delta/3$. For this x we have $\sigma(x) = \sigma(\xi)$.

Proof . For each $n \in \mathbf{N}$ we can select $x_n \in X$ so that $Qx_n = \xi$ and

$$\|nx_n\| \leq (1 + (1/n))\|n\xi\|.$$

For $m \geq n \geq 2$,

$$\begin{aligned} \|n(x_n - x_m)\| &\leq \|nx_n\| + \|mx_m\| \\ &\leq (2 + (1/n) + (1/m))\sigma(\xi) \\ &\leq \delta. \end{aligned}$$

However $n(x_n - x_m) \in B$ and so the set $\{n(x_n - x_m) : m \geq n \geq 2\}$ is bounded in X. Hence (x_n) is a Cauchy sequence in X. Let $x = \lim_{n\to\infty} x_n$. Then $Qx = \xi$ and if $m \in \mathbf{N}$,

$$\begin{aligned} \|mx\| &= \lim_{n\to\infty} \|mx_n\| \\ &\leq \lim_{n\to\infty} \|nx_n\| \\ &\leq \sigma(\xi). \end{aligned}$$

Thus $\sigma(x) = \sigma(\xi)$.

Finally if $Qy = \xi$ and $\sigma(y) \leq \delta/3$ then $x-y \in B$ and $\sigma(x-y) \leq (2/3)\delta$ so that $x = y$. □

LEMMA 6.3. Under the same assumptions as Lemma 6.2, suppose Y is a linear subspace of X/B with $\sigma(Y) \leq \delta/6$. Then there is a continuous linear map $h : Y \to X$ so that $Qh(\xi) = \xi$ for $\xi \in Y$.

Proof. Simply define $h(\xi)$ to be the unique $x \in X$ so that $Qx = \xi$ and $\sigma(x) \leq \delta/3$, as in Lemma 6.2. For $\alpha,\beta \in \mathbb{R}$, $\xi,\eta \in Y$,

$$\begin{aligned}\sigma(\alpha h(\xi)+\beta h(\eta)) &\leq \sigma(h(\xi)) + \sigma(h(\eta)) \\ &= \sigma(\xi) + \sigma(\eta) \\ &\leq \delta/3\end{aligned}$$

so that, from uniqueness,

$$h(\alpha\xi + \beta\eta) = \alpha h(\xi) + \beta h(\eta)$$

i.e., h is linear.

Suppose $\xi_n \in Y$ and $\|\xi_n\| \to 0$. Then we may select $x_n \in X$ with $Qx_n = \xi_n$ and

$$\|x_n\| \leq 2\|\xi_n\|.$$

Then $x_n-h(\xi_n) \in B$. To show h is continuous it will suffice to show that $\lim(x_n-h(\xi_n)) = 0$.

If not, by passing to a subsequence, we may suppose that for some $\alpha > 0$

$$\|\alpha(x_n-h(\xi_n))\| > \delta$$

(since the set $\{b \in B : \|b\| \leq \delta\}$ is bounded).

Then

$$\begin{aligned}\|\alpha x_n\| &\geq \delta - \|h(\alpha\xi_n)\| \\ &\geq \delta - \sigma(\xi_n) \\ &\geq 5\delta/6.\end{aligned}$$

However $\|x_n\| \to 0$ and so we have reached a contradiction. □

THEOREM 6.4. Let X be an F-space and let B be a closed locally bounded subspace of X. Let $T : L_0(0,1) \to X/B$ be a continuous linear operator. Then there is a unique linear operator $S : L_0(0,1) \to X$ so that $QS = T$, where $Q : X \to X/B$ is the quotient map.

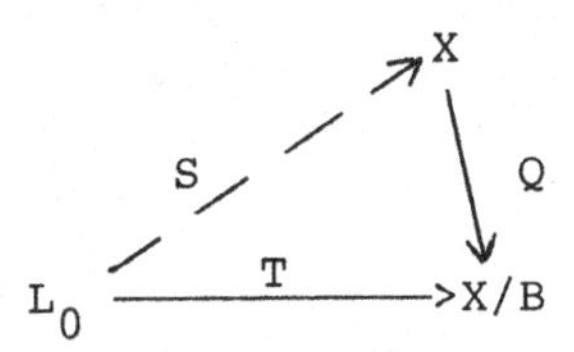

Proof. Choose $\delta > 0$ so that the set $\{b \in B : \|b\| \leq \delta\}$ is bounded. Then choose $\epsilon > 0$ so that if $f \in L_0$ and $\|f\| \leq \epsilon$ then $\|Tf\| \leq \delta/6$.

Partition $[0,1]$ into disjoint Borel sets $B_1,\ldots,B_N$ of measure less than ϵ and let $L_0(B_k)$ be the subspace of L_0 of all f supported on B_k for $1 \leq k \leq N$. For $1 \leq k \leq N$, let $Y_k = T(L_0(B_k))$.

If $f \in L_0(B_k)$

$$\sigma(f) = \sup_n \int_0^1 |nf(t)|/(1+|nf(t)|)dt$$
$$\leqslant \lambda(B_k) < \epsilon.$$

Hence $\sigma(Tf) \leqslant \delta/6$.

Now by Lemma 6.3 there exists a continuous linear map $h_k : Y_k \to X$ so that $Qh_k(\eta) = \eta$ for $\eta \in Y_k$. Define $S : L_0 \to X$ by

$$Sf = \sum_{k=1}^{N} h_k T(f \cdot 1_{B_k}).$$

It is clear that S is continuous and lifts T as required. Uniqueness follows from the fact that if S' similarly lifts T then $S-S'$ maps L_0 into B, and so $S-S' = 0$. □

THEOREM 6.5. L_0 is a K-space.

Proof. This is now an immediate deduction from Theorem 6.4 in view of Theorem 6.1. □

Theorems 6.4 and 6.5 can be generalized somewhat (cf. [Kalton-Peck (1979b)]) to apply to spaces of the type $L_0(X)$ where X is an F-space. $L_0(X)$ is the space of X-valued measurable functions equipped with the topology of convergence in measure. Drewnowski (private communication) has shown that Theorem 6.4 holds when $B \approx \omega$, the space of all real sequences. Of course ω is not locally bounded.

There must of course be some restriction on B in Theorem 6.4. Indeed L_0 is isomorphic to a quotient of an F-space with separating dual (Turpin [1976] prop. 0.3.11). This isomorphism cannot be lifted.

3. Lifting theorems for L_p when $0 < p < 1$.

In this section we give a very simple analogue of Theorem 6.4 for the spaces L_p for $0 < p < 1$. As we explain below this theorem is by no means the complete story, but will suffice for an application in the next section.

THEOREM 6.6. Suppose $0 < p < 1$ and that X is a p-Banach space. Suppose N is a closed subspace of X which is q-convex for some $q > p$. Let $T : L_p \to X/N$ be a bounded linear operator. Then there is a unique operator $S : L_p \to X$ so that $QS = T$ where $Q : X \to X/N$ is the quotient map.

REMARKS. Before giving the proof we offer some comments on this theorem. First note that if L_p is replaced by ℓ_p then no restriction on N is required to obtain a lifting as in the above theorem. The space ℓ_p is projective for the category of p-Banach spaces:

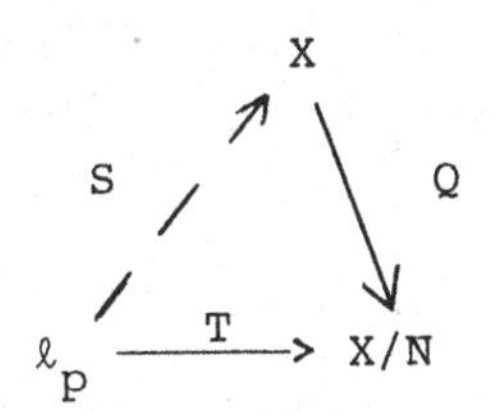

The proof of this is a simple application of the Open Mapping Theorem. Uniqueness, however, will not follow.

Replacing ℓ_p by L_p, which has the same local structure, does require some hypothesis on N however. In the case $p = 1$, it is essentially a result of Lindenstrauss [1964] that it suffices for N to be a dual Banach space. A similar hypothesis can be used for $0 < p < 1$; one can require that there exists on N a vector topology for which the unit ball is relatively compact. This is done in Kalton-Peck (1979b).

In fact Theorem 6.6 as stated is valid without the assumption that X is a p-Banach space; one can assume X is an r-Banach space for some $r \leq p$. The proof of this is however much more involved, depending on results from Kalton [1978b].

<u>Proof</u> <u>of</u> <u>Theorem</u> <u>6.6</u>. For each $n \in \mathbb{N}$ define $B_{n,k} = ((k-1)2^{-n}, k2^{-n}] \subset [0,1]$ for $1 \leq k \leq 2^n$ and let $\chi_{n,k}$ be the characteristic function of $B_{n,k}$. Let E_n be the linear span in $L_p(0,1)$ of $\{\chi_{n,k} : 1 \leq k \leq 2^n\}$.

As $\|\chi_{n,k}\| = 2^{-n/p}$ there is an $x_{n,k} \in X$ so that

$$Qx_{n,k} = Tx_{n,k},$$

$$\|x_{n,k}\| \leq 2^{1-n/p}\|T\|.$$

Define $S_n : E_n \to X$ by

$$S_n\left(\sum_{k=1}^{2^n} a_k\chi_{n,k}\right) = \sum_{k=1}^{2^n} a_k x_{n,k}$$

Then $\|S_n\| \leq 2\|T\|$, and $QS_nf = Tf$ for $f \in E_n$.

Since $\cup E_n$ is dense in L_p, it suffices to show that for $f \in \cup E_n$, $\lim S_nf$ exists. We then define S to be the unique extension of the operator $S_\infty : \cup E_n \to X$ defined by

$$S_\infty f = \lim_{n\to\infty} S_n f .$$

Uniqueness will then be automatic. If $S' : L_p \to X$ is another lift then $(S-S') : L_p \to N$ so that $S-S' = 0$. (See Kalton-Peck, 1979b.)

Consider $\chi_{j,k} \in E_j$ and suppose $m > n \geq j$. Then

$$S_m\chi_{j,k} - S_n\chi_{j,k} = (S_m - S_n) \sum_{i=1}^{2^{n-j}} \chi_{n,2^{n-j}k+i} .$$

Now N is q-convex and so there is a $C < \infty$ so that if $y_1,\ldots,y_\ell \in N$ then

$$\|y_1+\ldots+y_\ell\|^q \leq C^q \sum_{i=1}^{\ell} \|y_i\|^q .$$

Hence

$$\|S_m\chi_{j,k} - S_n\chi_{j,k}\|^q \leq C^q \|S_m - S_n\|^q 2^{n-j} 2^{-nq/p}$$

$$\leq C^q (\|S_m\|^p + \|S_n\|^p)^{q/p} \, 2^{n(1-q/p)-j}$$

$$\leq 2^{q+q/p-j} 2^{n(1-q/p)} C^q \|T\|^q .$$

Thus the sequence $\{S_n\chi_{j,k}\}_{n\geq j}$ is Cauchy and the proof is complete. □

4. Applications

For the purpose of this section let us define a subspace Y of L_p $(0 \leq p < 1)$ to be <u>liftable</u> if whenever we have an operator $T : L_p \to L_p/Y$ there is a unique lift $S : L_p \to L_p$ so that the following diagram commutes:

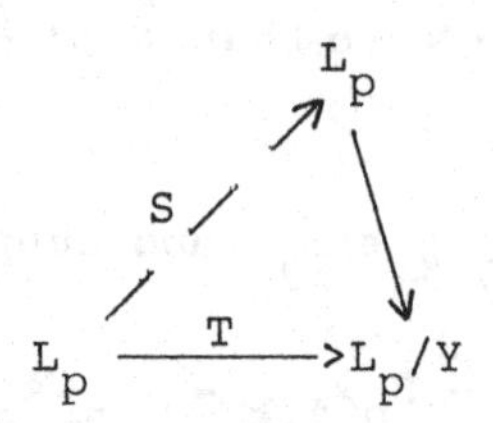

Clearly it is required for the lifting to be unique that are no nonzero operators from L_p into Y.

The results of Sections 2 and 3 give us numerous examples of liftable subspaces:

(i) Any locally bounded subspace of L_0, or a subspace isomorphic to ω, is liftable.

(ii) any subspace of L_p for $0 < p < 1$ which is q-convex for $q > p$, or has a vector topology for which the unit ball is relatively compact, is liftable.

In particular, finite-dimensional subspaces of L_p for $0 \leq p < 1$ are liftable.

THEOREM 6.7. Suppose $0 \leqslant p < 1$ and M and N are two subspaces of L_p such that L_p/M and L_p/N are isomorphic. If M and N are both liftable there is an automorphism $U : L_p \to L_p$ such that $U(M) = N$. In particular M and N are isomorphic.

Proof. Let $J : L_p/M \to L_p/N$ be an isomorphism. We have the following diagram, where Q_1 and Q_2 denote the natural quotient maps:

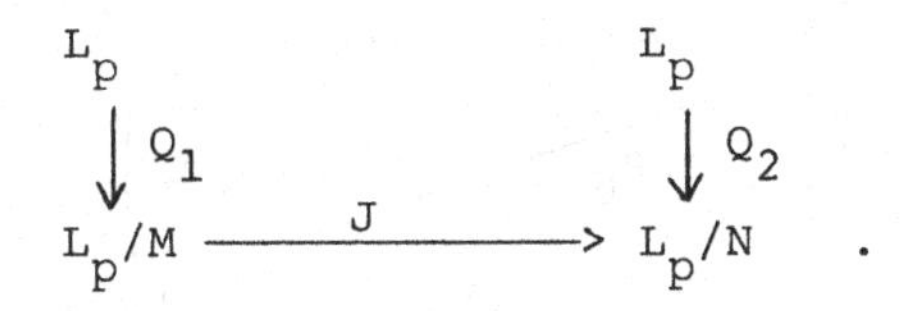

By the lifting property for N there is an operator $U : L_p \to L_p$ so that $Q_2U = JQ_1$. By the lifting property for M there is an operator $V : L_p \to L_p$ so that $Q_1V = J^{-1}Q_2$.

Now $Q_1VU = J^{-1}Q_2U = J^{-1}JQ_1 = Q_1$. Hence VU is a lift of the quotient map Q_1. By uniqueness $VU = I$. Similarly $UV = I$ so that U is indeed an automorphism.

Clearly $U(M) \subset N$ and $V(N) \subset M$ so that $U(M) = N$.

EXAMPLE. A non-liftable subspace of $L_p(T)$ with separating dual.

Consider the space J_p of Section 3.3. Since J_p is a subspace of H^p it has separating dual. However we claim that

J_p is not a liftable subspace. To see this we use Theorem 3.9. Consider the quotient map $A : H^p \oplus \overline{H}^p \to L_p(\mathbb{T})$ given by

$$A(f,g) = f + g$$

and the map $P : H^p \oplus \overline{H}^p \to H^p$ given by

$$P(f,g) = f.$$

If $(f,g) \in \ker A$, then $f \in J_p$. Hence P factors to a map $P_0 : L_p(\mathbb{T}) \to L_p/J_p$:

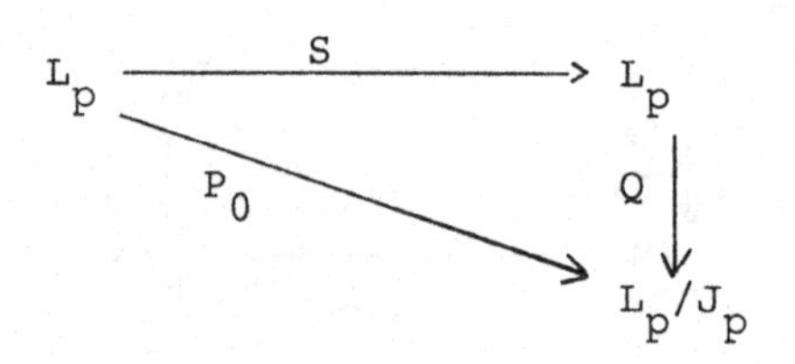

If J_p is liftable then there is an operator $S : L_p \to L_p$ so that the above diagram commutes. However the range of S, $\mathcal{R}(S) = \mathcal{R}(P) \subset H^p$ which has separating dual. Hence $S = 0$ which implies a contradiction.

We remark further that H^p is liftable; in fact it satisfies the second half of criterion (ii). Furthermore L_p/H^p and L_p/J_p are isomorphic (Aleksandrov [1981] or Kalton [to appear]), while H_p and J_p are not.

If $\dim M, \dim N < \infty$, Theorem 6.7 implies that if L_p/M and L_p/N are isomorphic then $\dim M = \dim N$. Our first aim in the next chapter will be to establish the converse. Note first that if $\dim M > 0$ then L_p/M and L_p are <u>not</u> isomorphic; contrast the situation when $p = 1$.

CHAPTER 7
TRANSITIVE SPACES AND SMALL OPERATORS

1. Introduction

In this chapter we shall deal only with quasi-Banach spaces. Some of the results can be stated in more generality but it is simplest to restrict ourselves throughout.

Let X and Y be quasi-Banach spaces. Then the space of operators $T : X \to Y$, denoted by $L(X,Y)$, is a quasi-Banach space when quasi-normed by

$$\|T\| = \sup(\|Tx\| : \|x\| \leq 1).$$

If $X = Y$ then we write $L(X,X) = L(X)$; $L(X)$ is an algebra.

An operator $K : X \to Y$ is said to be compact if the set $\{Kx : \|x\| \leq 1\}$ is relatively compact. K is strictly singular if for any closed subspace M of X such that $K|_M$ is an isomorphism, we have $\dim M < \infty$. Obviously any compact operator is also strictly singular.

The space X is said to be transitive if whenever $x,y \in X$ with $x \neq 0$ there exists $T \in L(X)$ with $Tx = y$.

The main theme of this chapter is that there is a surprising relationship between the notion of transitivity and the existence of "small"operators. Of course if X has a separating dual then it is automatically transitive (rank-one operators can be used in the definition). However we shall be much more interested in transitive spaces with trivial dual. As we shall see the spaces L_p for $0 < p < 1$ are examples.

Hyers [1939] and Williamson [1954] showed that the Fredholm theorem for compact operators on a Banach space goes through almost undisturbed without local convexity. As

Williamson observed, in a space with a trivial dual this leads to the conclusion that no compact endomorphism has a non-zero eigenvalue. This in turns leads to a theorem of Pallaschke [1973] that if X is a transitive space with trivial dual then every compact endomorphism $K : X \to X$ is identically zero. In particular this holds for the case $X = L_p$ for $0 < p < 1$ -- a result also obtained by different methods by Turpin [1973].

In this chapter we shall prove a previously unpublished generalization of Pallaschke's theorem to strictly singular operators.

In the light of Pallaschke's work Pelczynski was led to ask whether it is possible to have a non-zero compact endomorphism of any space with trivial dual. To answer this question it is enough to consider compact operators $K : X \to Y$ where X has trivial dual. For if $K : X \to Y$ is a bounded operator with dense range and $\varphi \in Y^*$, then $\varphi \circ K \in X^*$ i.e. $\varphi \circ K = 0$; hence $\varphi = 0$. Thus the operator $(x,y) \to (0,Kx)$ is an endomorphism of the space $X \oplus Y$, which has trivial dual.

Kalton and Shapiro [1975] gave an example to show that it is indeed possible to have a non-zero compact operator on a space with trivial dual. The example will be given in this chapter. However, in contrast, there is no non-zero strictly singular operator $T : L_p \to Y$ (for any range space Y), when $0 < p < 1$.

Our final example in this chapter is of a quasi-Banach space X whose algebra of operators $L(X)$ reduces simply to multiples of the identity operator I. Such a space is called rigid. The existence of rigid spaces had long been suspected when the first example was obtained by Roberts in 1977 (following the example of an incomplete rigid space of Waelbroeck [1977]).

2. Transitive spaces

A quasi-Banach algebra is an algebra A equipped with a quasi-norm so that it is complete and the multiplication satisfies

$$\|ab\| \leq \|a\|\cdot\|b\| \qquad a,b \in A.$$

Clearly if X is a quasi-Banach space then $L(X)$ is a quasi-Banach algebra.

In a series of papers in the 1960's Zelazko ([1960],[1965]) systematically studied quasi-Banach algebras and found that much of the theory of Banach algebras could be extended without tears. We shall have need of a couple of basic results on commutative quasi-Banach algebras.

THEOREM 7.1. (Zelazko [1960]). Let A be a commutative quasi-Banach algebra with identity 1. Define for $a \in A$

$$\rho(a) = \limsup_{n\to\infty} \|a^n\|^{1/n}.$$

Then ρ is a semi-norm on A, $\rho(1) = 1$ and

$$\rho(ab) \leq \rho(a)\rho(b) \qquad a,b \in A.$$

Proof. All that requires proof is that ρ is a semi-norm. We note that by the Aoki-Rolewicz theorem, for some p, $0 < p < 1$ and $C < \infty$ we have

$$\|a_1+\ldots+a_n\|^p \leq C(\sum_{i=1}^n \|a_i\|^p)^{1/p}$$

for $a_1,\ldots,a_n \in A$.

Now let $a,b \in A$ and suppose $\gamma > \rho(a) + \rho(b)$. Then there is a constant $M < \infty$ so that

$$\|a^n\| \leq M(\gamma\rho(a)/(\rho(a)+\rho(b)))^n \qquad n = 1,2,\ldots$$

$$\|b^n\| \leq M(\gamma\rho(b)/(\rho(a)+\rho(b)))^n \qquad n = 1,2,\ldots$$

Now

$$(a+b)^n = \sum_{k=0}^{n} \binom{n}{k} a^k b^{n-k}$$

so that

$$\|(a+b)^n\| \leq CM^2(\gamma/(\rho(a)+\rho(b)))^n \left(\sum_{k=0}^{n} \binom{n}{k}^p \rho(a)^{kp}\rho(b)^{(n-k)p}\right)^{1/p}$$
$$\leq CM^2\gamma^n n^{1/p}.$$

Hence $\rho(a+b) \leq \gamma$. As $\gamma > \rho(a) + \rho(b)$ is arbitrary we have $\rho(a+b) \leq \rho(a) + \rho(b)$. The other semi-norm properties are obvious. □

THEOREM 7.2. (Gelfand-Mazur-Zelazko) Let A be a quasi-Banach field. Then

(i) If A is complex, $A \approx C$.
(ii) If A is real, $A \approx R$ or $A \approx C$.

<u>Proof</u>. Observe that the set $\{a \in A : \rho(a) = 0\}$ is an ideal. Hence ρ is a norm on A and (A,ρ) is a normed field. The result follows by the standard Gelfand-Mazur theorem. □

[We remark here that completeness is normally assumed in the Gelfand-Mazur theorem, but it is not required. For example, in the complex case, one may obtain the proof by considering any non-trivial multiplicative linear functional on the completion of (A,ρ).]

If X is a quasi-Banach space, then we define $A(X)$ to be the center of $L(X)$ i.e. the set of $S \in L(X)$ such that $ST = TS$ for every $T \in L(X)$. $A(X)$ is a commutative quasi-Banach algebra. It is possible (as we shall see) for $L(X)$ to be commutative so that $L(X) = A(X)$; however for transitive spaces $A(X)$ is small.

THEOREM 7.3. Let X be a transitive quasi-Banach space. Then $A(X)$ is a field, and hence:

(i) If X is complex, $A(X) = (\lambda I : \lambda \in \mathbf{C})$

(ii) If X is real, either $A(X) = (\lambda I : \lambda \in \mathbf{R})$ or there is an operator $J \in A(X)$ with $J^2 = -I$ so that $A(X) = (\lambda I + \mu J : \lambda, \mu \in \mathbf{R})$.

Proof. If $S \in A(X)$ then $\ker S$ and $S(X)$ are invariant subspaces for every $T \in L(X)$. Since X is transitive either $S = 0$ or S is invertible. The theorem immediately follows. □

PROBLEM. Give an example of a transitive real quasi-Banach space such that $A(X) \approx \mathbf{C}$. Such a space would admit a natural "complex structure" (let $J = i$), such that every operator is automatically complex-linear. It would admit a second complex structure (let $J = -i$) so that the resulting complex quasi-Banach spaces are non-isomorphic.

DEFINITION. A transitive real quasi-Banach space X is strictly transitive if $A(X) = \{\lambda I : \lambda \in \mathbf{R}\}$.

THEOREM 7.4. The real spaces $L_p(0,1)$ for $0 < p < 1$ are strictly transitive. The complex spaces $L_p(0,1)$ are transitive.

Proof. We shall give the proof only in the real case. First we prove (cf. Rolewicz [1972] pp. 253-254) that L_p is transitive. Indeed we shall show that if $f,g \in L_p$ with $\|f\| = \|g\| = 1$ there exists $T \in L(L_p)$ with $\|T\| = 1$ and $Tf = g$.

Let μ_f and μ_g be the probability measures on $(0,1)$ given by

$$\mu_f(B) = \int_B |f|^p d\lambda \qquad B \in \mathcal{B}$$

$$\mu_g(B) = \int_B |g|^p d\lambda \qquad B \in \mathcal{B}$$

(where $\mathcal{B}$ is the class of Borel subsets of $(0,1)$). Then since μ_f and μ_g are both nonatomic there is a Borel map $\sigma : (0,1) \to (0,1)$ (cf. Royden [1963], Chapter 14) so that

$$\mu_g(B) = \mu_f(\sigma^{-1}B) \qquad B \in \mathcal{B}$$

Define $T : L_p \to L_p$ by

$$Th(s) = g(s)f(\sigma s)^{-1}h(\sigma s)$$

when $f(\sigma s) \neq 0$ and

$$Th(s) = 0$$

otherwise. Now

$$\begin{aligned}
\|Th\|^p &= \int_{|f(\sigma s)|>0} |g(s)|^p(|h(\sigma s)|^p)/(|f(\sigma s)|^p)\, d\lambda(s) \\
&= \int_{|f(\sigma s)|>0} (|h(\sigma s)|^p)/(|f(\sigma s)|^p)\, d\mu_g(s) \\
&= \int_{|f(s)|>0} (|h(s)|^p)/(|f(s)|^p)\, d\mu_f(s) \\
&= \int_{|f(s)|>0} |h(s)|^p d\lambda(s) \\
&\leq \|h\|^p
\end{aligned}$$

so that $\|T\| \leq 1$.

Now

$$\begin{aligned}
Tf(s) &= g(s) \qquad & f(\sigma s) \neq 0 \\
&= 0 \qquad & f(\sigma s) = 0.
\end{aligned}$$

However

$$\int_{f(\sigma s)=0} |g(s)|^p d\lambda(s) = \mu_g\{s : f(\sigma s) = 0\}$$

$$= \mu_f\{s : f(s) = 0\}$$

$$= \int_{|f(s)|=0} |f(s)|^p d\lambda(s)$$

$$= 0.$$

Hence $Tf = g$.

Next we consider $A(L_p)$. For $f \in L_\infty(0,1)$ let $M_f \in L(L_p)$ be defined by

$$M_f(g) = f \cdot g.$$

For $s \in (0,1)$ let $T_s \in L(L_p)$ be defined by

$$T_s(f)[t] = f(st).$$

Note that $\|M_f\| = \|f\|_\infty$ and $\|T_s\| = s^{-1/p}$.

Suppose $S \in A(L_p)$ and $S1 = \varphi \in L_p$. Then for $f \in L_\infty$

$$Sf = SM_f 1 = M_f \varphi = f \cdot \varphi.$$

Hence $S = M_\varphi$. For $0 < s < 1$

$$\varphi = S1$$

$$= ST_s 1$$

$$= T_s S1$$

$$= T_s \varphi.$$

Hence

$$\varphi(t) = \varphi(st) \qquad t \text{ a.e., } \quad 0 < s < 1.$$

By Fubini's theorem, for almost every t,

$$\varphi(t) = \varphi(st) \qquad s \text{ a.e.}$$

Hence φ is a constant almost everywhere, and $S = \lambda I$ for some $\lambda \in \mathbb{R}$. Thus L_p is strictly transitive. □

THEOREM 7.5. Suppose X is a real strictly transitive quasi-Banach or any complex transitive quasi-Banach space. Then if $\{x_1,\ldots,x_n\}$ is a linearly independent set and $\{y_1,\ldots,y_n\} \subset X$, there exists $T \in L(X)$ with $Tx_i = y_i$.

Proof. Clearly it is enough to show that if $\{x_1,\ldots,x_n\}$ is linearly independent and $u \in X$ then there exists $T \in L(X)$ with $Tx_1 = Tx_2 = \ldots = Tx_{n-1} = 0$ and $Tx_n = u$. Again, since X is transitive, it suffices to show $Tx_n \neq 0$ for some $T \in L(X)$ such that $Tx_i = 0$, $1 \leq i \leq n-1$.

We first prove this for $n = 2$. If it is not true then $Tx_1 = 0$ implies $Tx_2 = 0$ for $T \in L(X)$. From this we deduce that there is a linear map $S : X \to X$ so that

$$Tx_2 = STx_1 \qquad T \in L(X).$$

Clearly $Sx_1 = x_2$ and if $x \in X$ there exists $V \in L(X)$ with $Vx_1 = x$. Hence

$$\begin{aligned} STx &= STVx_1 \\ &= TVx_2 \\ &= TSVx_1 \\ &= TSx. \end{aligned}$$

Thus $TS = ST$ for $T \in L(X)$. Next we show $S \in L(X)$. Suppose $u_n \to 0$. The map $T \to Tx_1$ $(L(X) \to X)$ is open and hence there exist $T_n \in L(X)$ with $\|T_n\| \to 0$ and

$$u_n = T_n x_1.$$

Now

$$Su_n = ST_n x_1$$

$$= T_n x_2.$$

Hence $\|Su_n\| \to 0$ and $S \in L(X)$.

We conclude $S \in A(X)$ and hence $S = \lambda I$ for some $\lambda \in \mathbf{R}$ (or $\mathbf{C}$). This contradicts the linear independence of x_1 and x_2.

We now prove the general case by induction. Suppose the result is known for $n < m$, and that $\{x_1, \ldots, x_m\}$ is linearly independent. If the theorem is false for this collection then we may assume that $Tx_1 = \ldots = Tx_{m-1} = 0$ implies $Tx_m = 0$. Now if $T \in L(X)$ and $Tx_1 = \ldots = Tx_{m-2} = 0$ and $\{Tx_{m-1}, Tx_m\}$ are linearly independent, there exists $S \in L(X)$ with $STx_1 = \ldots = STx_{m-1} = 0$ but $STx_m \neq 0$. Hence for every such T, $\{Tx_{m-1}, Tx_m\}$ is linearly dependent. It quickly follows that for some constants α, β not both zero we have

$$\alpha Tx_{m-1} + \beta Tx_m = 0$$

for every such T. But then $Tx_1 = \ldots = Tx_{m-2} = 0$ implies

$T(\alpha x_{m-1} + \beta x_m) = 0$, contradicting the inductive hypothesis applied to $\{x_1,\ldots,x_{m-2},\alpha x_{m-1} + \beta x_m\}$. □

THEOREM 7.6. Let X be a real strictly transitive quasi-Banach space, or a complex transitive quasi-Banach space. Let V and W be two n-dimensional subspaces of X. Then there exists an automorphism T of X so that $T(V) = W$.

Proof. Let B_n be the set of all linearly independent sets in X^n. It is easy to show that B_n is connected. Indeed given any two members of B_n one can construct a path between them by restricting attention to a subspace of X of dimension at most 2n.

Define an equivalence relation $\approx$ on B_n by $(u_1,\ldots,u_n) \approx (v_1,\ldots,v_n)$ if there is an automorphism $T \in L(X)$ with $Tu_i = v_i$. The equivalence classes are open sets since if $(u_1,\ldots,u_n) \in B_n$ there exists by the Open Mapping Theorem and Theorem 7.5, $\delta > 0$ such that if $\|x_i\| < \delta$ $(1 \leq i \leq n)$ there is $S \in L(X)$ with $\|S\| < 1$ and $Su_i = x_i$. But then $I + S$ is an automorphism of X and

$$(I+S)u_i = u_i + x_i \qquad 1 \leq i \leq n.$$

Hence there is only one equivalence class and the theorem is proved. □

COROLLARY 7.7. Let V and W be two finite-dimensional subspaces of L_p for $0 < p < 1$. Then $L_p/V \approx L_p/W$ if and only if $\dim V = \dim W$.

Proof. This follows from Theorem 6.7. □

REMARK. This also holds when $p = 0$ - see Kalton-Peck [1979b]. In this same paper it was shown that there are two subspaces V and W of L_0, each isomorphic to ℓ_2, while L_0/V is not isomorphic to L_0/W. Peck and Starbird [1981] have shown that $L_0/V \approx L_0/W$ whenever V and W are both isomorphic to ω. It is unknown at present for $0 < p < 1$ whether $V \approx W \approx \ell_p$ implies $L_p/V \approx L_p/W$. For $p = 1$, however, this is known to be false by a recent result of Bourgain [1981].

3. Strictly singular endomorphisms

DEFINITION. An operator $T \in L(X,Y)$ is a Fredholm operator if $T(X)$ is closed and $\ker T$ and $\operatorname{coker} T = Y/T(X)$ are finite-dimensional.

T is semi-Fredholm if $\ker T$ is finite-dimensional and $T(X)$ is closed.

EXAMPLE. Suppose $T \in L(L_p)$ is Fredholm, when $0 < p < 1$. Then $T(X)$ is of finite co-dimension and hence as $L_p^* = \{0\}$, $T(X) = L_p$ i.e. T is onto. Let $V = \ker T$. Then $L_p/V \approx L_p$. Hence (Theorem 7.6, Corollary 7.7) $V = \{0\}$ i.e. T is invertible.

PROPOSITION 7.8. In $L(X,Y)$ let G_1 denote the set of isomorphic embeddings and G_2 the set of open mappings. Then G_1 and G_2 are open sets.

Proof. If $S \in G_1$ then for some $\alpha > 0$,

$$\|Sx\| \geqslant \alpha\|x\| \qquad x \in X.$$

Thus if $\|T\| < \alpha$, $S+T \in G_1$, i.e., G_1 is open.

If $S \in G_2$ then for some $\alpha > 0$, given $y \in Y$ there exists $x \in X$ with

$$Sx = y$$

$$\|x\| \leqslant \alpha\|y\|.$$

Suppose $\|T\| < \alpha^{-1}$. We show $S+T \in G_2$, i.e. $S+T$ is onto. Fix $y \in Y$ and pick x_1 with $Sx_1 = y$ but $\|x_1\| \leqslant \alpha\|y\|$. Then

$$\|y-(S+T)x_1\| \leqslant \alpha \ \|T\| \|y\|$$

Let $y_1 = y - (S+T)x_1$ and pick x_2 so that

$$Sx_2 = y_1$$

$$\|x_2\| \leqslant \ \|y_1\|.$$

Proceeding by induction we obtain sequences (x_n), (y_n) so that

$$Sx_n = y_{n-1}, \qquad n \geqslant 1$$

(with $y_0 = y$)

$$\|x_n\| \leqslant \alpha\|y_{n-1}\| ,$$

$$y_n = y_{n-1} - (S+T)x_n$$

$$= -Tx_n.$$

Hence

$$\|y_n\| \leq (\alpha\|T\|)^n \|y_0\|$$

and

$$\|x_n\| \leq \alpha(\alpha\|T\|)^{n-1}\|y_0\|.$$

Let $x = \sum_{n=1}^{\infty} x_n$. Then

$$(S+T)x = \sum_{n=1}^{\infty}(y_{n-1}-y_n) = y.$$

Thus $S+T \in G_2$. □

PROPOSITION 7.9. $G_1 \cap G_2$ is a clopen subset of both G_1 and G_2.

Proof. We may suppose the quasi-norm p-subadditive where $p < 1$. First consider $G_1 \cap G_2 \subset G_1$. Clearly $G_1 \cap G_2$ is open relative to G_1. Now suppose $S_n \in G_1 \cap G_2$ and $\|S_n - S\| \to 0$ where $S \in G_1$. Let $S(X) = E$; E is a closed subspace of Y and there is an operator $R \in L(E,X)$ with $RS = I_X$ and $SR = I_E$.

For $e \in E$

$$S_n^{-1}e = RSS_n^{-1}e$$

$$= Re + R(S-S_n)S_n^{-1}e.$$

Let $\alpha_n = \sup(\|S_n^{-1}e\| : \|e\| \leq 1,\ e \in E)$. Then

$$\alpha_n^p \leq \|R\|^p (1+\|S-S_n\|^p \alpha_n^p).$$

Since $\|S-S_n\| \to 0$ we conclude

$$\sup \alpha_n = \alpha < \infty.$$

Hence

$$\|S_n^{-1}e - Re\| = \|R(S-S_n)S_n^{-1}e\|$$
$$\leqslant \alpha\|R\| \|S-S_n\| \|e\|.$$

Thus $S_n^{-1} \to R$ in $L(E,X)$. Hence as $R \in G_2(E,X)$, $S_n^{-1} \in G_2(E,X)$ eventually, i.e. $S_n^{-1}(E) = X$ for large n. Since $S_n \in G_1 \cap G_2$ this means that $E = Y$ and $S \in G_2$.

Next consider $G_1 \cap G_2 \subset G_2$; again it is an open subset. Suppose $S_n \in G_1 \cap G_2$, $\|S_n - S\| \to 0$ where $S \in G_2$. Hence for some $\alpha < \infty$, if $y \in Y$ there exists $x \in X$ with $\|x\| \leqslant \alpha\|y\|$ and $Sx = y$.

Now

$$S_n^{-1}y = S_n^{-1}Sx$$
$$= x + S_n^{-1}(S-S_n)x.$$

Hence $\|S_n^{-1}\|^p \leqslant \alpha^p(1+\|S_n^{-1}\|^p\|S-S_n\|^p)$. Thus

$$\sup\|S_n^{-1}\| = M < \infty$$

and so if $x \in X$

$$\|x\| \leqslant M\|S_n x\| \qquad n = 1,2,\ldots$$

so that

$$\|x\| \leqslant M\|Sx\|$$

and $S \in G_1$. □

THEOREM 7.10. Let $J \in G_1(X,Y)$ and suppose $T \in L(X,Y)$ is strictly singular; then $J+T$ is semi-Fredholm.

Proof. On $N = \ker(J+T)$, $T = -J$ is an isomorphism. Hence $\dim N < \infty$. Let $q : X \to X/N$ be the quotient mapping and let $A \in L(X/N,Y)$ be such that $Aq = J+T$. Then $(A) = (J+T)$ and we must show this is closed, i.e. A is an isomorphism.

Suppose not. Then the quasi-norm $\xi \to \|A\xi\|$ on X/N induces a weaker Hausdorff vector topology on X/N. It is possible thus to find a sequence $\xi_n \in X/N$ with $\|\xi_n\| = 1$ and $\|A\xi_n\| < 2^{-n}$. By passing to a subsequence and using Theorem 4.7, we may suppose (ξ_n) is a strongly regular M-basic sequence. Let $E = [\xi_n]_{n=1}^{\infty}$ and $E_k = [\xi_n]_{n \geq k}$ for $k \geq 2$. Let $\xi_n^* \in E^*$ be such that

$$\xi_i^*(\xi_j) = \delta_{ij}$$

$$\|\xi_i^*\| \leq K \qquad i = 1,2,\ldots$$

for some $K < \infty$.

If $\eta \in E_k$,

$$A(\eta) = \sum_{i=k}^{\infty} \xi_i^*(\eta)A\xi_i$$

so that

$$\|A(\eta)\|^p \leq CK^p\|\eta\|^p 2^{-k},$$

for some C independent of k.

Now for some $\lambda < \infty$, we have

$$\|Jx\| \geq \lambda\|x\| \qquad x \in X.$$

For large enough k

$$\|A\eta\| \leqslant (1/2)\lambda \|\eta\| \qquad \eta \in E_k,$$

so that if $x \in q^{-1}E_k$,

$$\|(J+T)x\| \leqslant (1/2)\lambda \|x\|.$$

This means that T is an isomorphism on $q^{-1}E_k$ and we have a contradiction. □

PROPOSITION 7.11. Suppose X is a transitive quasi-Banach space with trivial dual. Suppose $T \in L(X)$ is semi-Fredholm. Then if

$$\Lambda_T(A) = TA \qquad A \in L(X)$$

then $\Lambda_T \in G_1(L(X), L(X))$.

Proof. If $\Lambda_T(A) = 0$ then $TA = 0$ i.e. $R(A) \subset \ker T$. As $X^* = \{0\}$ and $\dim(\ker T) < \infty$ this means $A = 0$.

Now suppose $\|TA_n\| \to 0$ but $\|A_n\| = 1$. Pick $x_n \in X$ with $\|x_n\| = 1$ and $\|A_n x_n\| \geqslant 1/2$. Fix $u \in X$ with $\|u\| = 1$. By the Open Mapping Theorem, for some $\lambda < \infty$ we can find $B_n \in L(X)$ with

$$B_n u = x_n$$

$$\|B_n\| \leqslant \lambda.$$

Now

$$\|TA_nB_n\| \leqslant \lambda\|TA_n\| \to 0.$$

For each $x \in X$, $\{A_nB_nx\}$ is bounded and $TA_nB_nx \to 0$. As T is semi-Fredholm there exist $u_n \in \ker T$ with

$$\|A_n B_n x - u_n\| \to 0$$

and $\{u_n : n = 1,2,\ldots\}$ is relatively compact. Hence $\{A_n B_n x : n \in \mathbf{N}\}$ is relatively compact.

Let $\mathcal{U}$ be a free ultrafilter on $\mathbf{N}$ and define $S \in \mathcal{L}(X)$ by

$$Sx = \lim_{n \in \mathcal{U}} A_n B_n x.$$

Then $\|S\| \leqslant \lambda$ and $\mathcal{R}(S) \subset \ker T$. Hence $S = 0$.

As this is true for every ultrafilter $\mathcal{U}$, we conclude

$$\lim_{n \to \infty} \|A_n B_n x\| = 0 \qquad x \in X$$

contrary to assumption, since $A_n B_n u = A_n x_n$. Hence $\Lambda_T \in G_1$ as required. □

THEOREM 7.12. Let X be a transitive quasi-Banach space with trivial dual. If $T \in \mathcal{L}(X)$ is strictly singular then $T = 0$.

Proof. Consider for $\alpha \in \mathbf{R}$ (or $\mathbf{C}$), $\Lambda_{I+\alpha T} : \mathcal{L}(X) \to \mathcal{L}(X)$. Putting together Theorems 7.10 and 7.11, $\Lambda_{I+\alpha T} \in G_1$ for every α. Hence the set $\{\alpha \in \mathbf{R}: \Lambda_{I+\alpha T} \in G_1 \cap G_2\}$ is clopen and contains 0. Thus $\Lambda_{I+\alpha T}$ is invertible for all α, and so $I+\alpha T$ is invertible for all α.

If $T \neq 0$ there exists $x \in X$ with $Tx \neq 0$. Since X is transitive there exists $S \in \mathcal{L}(X)$ with $STx = x$. Now $I-ST$ is not invertible and the above argument applied to ST in place of T produces a contradiction. □

REMARKS. This theorem generalizes Pallaschke's results and shows that L_p $(0 < p < 1)$ has no strictly singular endomorphisms. We shall shortly improve on this result.

If we consider compact rather than strictly singular operators we can get away with weaker assumptions. The next result is due to Williamson [1954].

THEOREM 7.13. Suppose X has trivial dual and $K \in \mathcal{L}(X)$ is compact. Then K has no non-zero eigenvalues.

Proof. Let $\mathcal{A}_K$ be the set of $S \in \mathcal{L}(X)$ which commute with K. We show that for every $\alpha \in \mathbf{R}$ (or $\mathbf{C}$), $\Lambda_{I+\alpha K} : \mathcal{A}_K \to \mathcal{A}_K$ is in the class G_1.

Arguing as in Proposition 7.11, $\ker \Lambda_{I+\alpha K} = 0$. Suppose $\|A_n\| = 1$ but $\|(I+\alpha K)A_n\| \to 0$. Using the fact that $I+\alpha K$ is semi-Fredholm, an argument similar to that of Proposition 7.11 shows that $A_n x \to 0$ for every $x \in X$. Hence $\|A_n K\| \to 0$ and so $\|A_n(I+\alpha K)\| \to 1$. Since A_n commutes with $I+\alpha K$ we have a contradiction.

Just as in Theorem 7.12 we conclude that $\Lambda_{I+\alpha K} \in G_1 \cap G_2$ for all α and hence $I+\alpha K$ is invertible for all α, which yields the theorem. □

COROLLARY 7.14. (Pallaschke) Suppose X has trivial dual and there is a dense subspace X_0 of X such that if $u \in X_0$ and $x \in X$ with $x \neq 0$ then there exists $T \in \mathcal{L}(X)$ with $Tx = u$. Then X has no non-trivial compact endomorphisms.

This Corollary applies to Orlicz function spaces (taking $X_0 = L_\infty$). See also Turpin [1973]. Orlicz function spaces in general fail to be transitive. (Kalton [1978c]).

4. Compact operators on spaces with trivial dual

In this section we describe an example of a compact operator whose domain has trivial dual, and hence a compact endomorphism of a space with trivial dual. This example is due to Kalton and Shapiro [1975]. We also construct a non-locally convex space on which the compact operators are sufficient to distinguish points from closed subspaces.

Our example depends on the notion of a mixed topology and in particular on some results of Wiweger [1961] and Waelbroeck [1971].

Let us suppose X is a quasi-Banach space, where we may assume the quasi-norm r-subadditive for some r, $0 < r \leq 1$. Let α be any vector topology on X which is weaker than the usual topology, but is not assumed Hausdorff. Let $\mathcal{V}$ be a base of closed balanced neighborhoods of zero for α. For every sequence (V_n) from $\mathcal{V}$ and sequence $p_n \in \mathbb{R}$ with $0 < p_n \uparrow \infty$ let

$$W(\{V_n\});\{p_n\}) = \bigcup_{N=1}^{\infty} \sum_{n=1}^{N} p_n(V_n \cap B)$$

where B is the closed unit ball of X. Now, as shown by Wiweger, the class of sets $W(\{V_n\};\{p_n\})$ is the base at 0 for some vector topology $\hat{\alpha}$ on X. Direct calculation shows that $\hat{\alpha} \geq \alpha$, and that $\hat{\alpha}$ is stronger than a vector topology β on X, whenever $\beta \leq \alpha$ on B. Thus we have:

PROPOSITION 7.15. $\hat{\alpha}$ is the finest vector topology on X agreeing with α on every bounded set.

PROPOSITION 7.16. Suppose X is p-convex; then $\hat{\alpha}$ has a base at 0 of q-convex neighborhoods for every $q < p$.

Proof. We can and do assume that B is p-convex. It will suffice to show that if $V \in \mathcal{V}$ then there exists $U \in \mathcal{V}$ with $co_q(U \cap B) \subset V$. Then it is easy to check from the definition that $\hat{\alpha}$ is locally q-convex.

Choose $W \in \mathcal{V}$ so that $W + W \subset V$ and then $\epsilon > 0$ so that $\epsilon B \subset W$. Pick $N \in \mathbf{N}$ so that $N > \epsilon^{-qp/p-q}$ and $U \in \mathcal{V}$ so that $U + U + \ldots + U((N+1)$ times$) \subset W$. Suppose $w \in co_q(U \cap B)$. Then

$$w = c_1u_1 + \ldots + c_nu_n$$

where $u_i \in U \cap B$ and $c_i \geq 0$, $c_1^q + \ldots + c_n^q = 1$. We can assume $c_1 \geq c_2 \ldots \geq c_n \geq 0$, and that $n \geq N$. Then

$$c_N^q \leq N^{-1}$$

and hence

$$\begin{aligned} c_N^p + \ldots + c_n^p &\leq c_N^{p-q}(c_N^q + \ldots + c_n^q) \\ &\leq c_N^{p-q} \\ &\leq N^{1-p/q}. \end{aligned}$$

Thus

$$\|c_Nu_N + \ldots + c_nu_n\| \leq N^{1/p - 1/q} < \epsilon$$

so that $c_Nu_N + \ldots + c_nu_n \in \epsilon B \subset W$.

Now $c_1u_1 + \ldots + c_{N-1}u_{N-1} \in U + \ldots + U \subset W$ so that $w \in W + W \subset V$, as required. □

THEOREM 7.17. (Waelbroeck [1971]). Suppose now that α is a Hausdorff topology and that B is α-compact. Then a set $C \subset X$ is $\hat{\alpha}$-closed if and only if $C \cap mB$ is α-closed for every $m \in \mathbf{N}$.

Proof. Let α^* be the finest topology on X which agrees with α on each set mB, $m \in \mathbf{N}$. Thus a set C is α^*-closed if and only if $C \cap mB$ is α-closed for every $m \in \mathbf{N}$. We need to show that $\alpha^* = \hat{\alpha}$. For this we need that α^* is a vector topology; then of course since α^* agrees with α on B we have $\alpha^* = \alpha$.

To achieve this we define yet another vector topology β. For each real sequence p_n, $0 \leq p_n \uparrow \infty$ and each $(V_n) \subset \mathcal{V}$, let

$$W(\{V_n\};\{p_n\}) = \bigcap_{n=1}^{\infty} (p_n B+V_n).$$

These sets form a base for a vector topology β. [In fact $\beta = \alpha$, see Wiweger [1961], but we do not need this.]

If $m \in \mathbf{N}$,

$$\overline{W}(\{V_n\};\{p_n\}) \cap mB = mB \cap \bigcap_{p_n \leq m} (p_n B+V_n)$$

is an α-neighborhood of 0 relative to mB. Hence $\beta \leq \alpha \leq \alpha^*$.

The topology α^* is translation-invariant since the maps $\tau_u : (X,\alpha^*) \to (X,\alpha^*)$ given by $\tau_u(x) = x+u$ are α-continuous on bounded sets. Hence to show that $\alpha^* \leq \beta$ requires us only to consider a base at the origin. Let U be any α^*-open set containing 0. We shall choose a sequence (V_n) from $\mathcal{V}$ by induction so that for each $n \in \mathbf{N}$

$$nB \cap \bigcap_{k=1}^{n} [(k-1)B+V_k] \subset U.$$

Indeed to start the process choose $V_1 \in \mathcal{V}$ so that $B \cap V_1 \subset U$; this is possible since 0 is α-interior to U relative to B. Now suppose $V_1,\ldots,V_m$ have been chosen. Suppose we <u>cannot</u> choose V_{m+1} so that the inductive hypothesis holds for $n = (m+1)$. Then for each $V \in \mathcal{V}$ we have

$$(m+1)B \cap \bigcap_{k=1}^{m}[(k-1)B+V_k] \cap (mB+V) \cap (X \setminus U) \neq \emptyset.$$

Note that $(m+1)B$ is α-compact. Each of the other sets in the intersection is closed (A_1+A_2 is closed if one is compact and the other is closed.)

Hence by the finite intersection property,

$$(m+1)B \cap \bigcap_{k=1}^{m} [(k-1)B+V_k] \cap \bigcap_{V\in} (mB+V) \cap (X \setminus U) \neq \emptyset.$$

However

$$\bigcap_{V\in\mathcal{V}} (mB+V) = mB$$

and so

$$mB \cap \bigcap_{k=1}^{m} [(k-1)B+V_k] \cap (X \setminus U) \neq \varphi$$

contrary to the inductive hypothesis.

Thus we can select (V_n) as required, and then we will have

$$\overline{W(\{V_n\};\{n-1\})} \subset U$$

so that $\alpha^* \leqslant \beta$. Then $\beta = \alpha = \alpha^*$. □

REMARK. Compare the Banach-Dieudonne theorem (Kelley-Namioka [1963] p. 211-212), which is a special case.

THEOREM 7.18. (Kalton-Shapiro) Let S_μ be a singular inner function on the unit disc in the complex plane (see Chapter 3). For $0 < p < 1$, there exists a non-zero compact operator $K : H^p/S_\mu H^p \to Y$ where Y is some quasi-Banach space. In particular if $S_\mu H^p$ is weakly dense, then $H^p/S_\mu H^p$ has trivial dual and admits compact operators.

Proof. Choose for α the topology on H^p of uniform convergence on compact subsets of the open unit disc D. The unit ball B of H^p is α-compact. Indeed if $f \in B$ then (see Chapter 3, or Duren [1970] p. 36),

$$|f(z)| \leq 2^{1/p}(1-|z|)^{-1/p} \qquad z \in D.$$

Hence B is a normal family of analytic functions and every sequence (f_n) in B has a convergent subsequence for α. Of course the limit of a convergent subsequence is still in B.

Thus we can apply Theorem 7.15 to α to produce the topology $\hat{\alpha} = \alpha^*$. We check $S_\mu H^p$ is $\hat{\alpha}$-closed. We must show $S_\mu H^p \cap mB$ is α-closed for every $m \in \mathbf{N}$. Indeed suppose $f_n \in S_\mu H^p \cap mB$ and $f_n \to f$ in α. Then $f_n = S_\mu g_n$ and $|g_n(z)| = |f_n(z)|$ a.e. for $|z| = 1$. Hence $\|g_n\| \leq m$ and g_n has a convergent subsequence g_{n_k} in the topology α. For $|z| < 1$,

$$\lim_{k\to\infty} S_\mu(z)g_{n_k}(z) = \lim_{k\to\infty} f_{n_k}(z) = f(z).$$

Hence if g is a limit point of g_{n_k}, we must have $S_\mu g = f$ so that $f \in S_\mu H^p \cap mB$.

Now in the quotient $\hat{\alpha}$-topology $H^p/S_\mu H^p$ is a Hausdorff space and its unit ball is compact for this topology. The quotient $\hat{\alpha}$-topology is locally q-convex for every $q < p$. Let Y be the completion of $H^p/S_\mu H^p$ for some non-trivial q-semi-norm which is $\hat{\alpha}$-continuous. Then the identity map $J : H^p/S_\mu H^p \to Y$ is compact. □

REMARK. The real point here is to establish the existence of a compact operator $K : H^p \to Y$ whose kernel is a PCWD-subspace. The most general result along these lines is in [1978e]:

THEOREM 7.19. Let X be a non-locally convex quasi-Banach space which is isomorphic to a subspace of a quasi-Banach space with a basis. Then there is a compact operator $K : X \to Y$ (for some quasi-Banach space Y) whose kernel is a PCWD-subspace.

Some restriction on X is necessary here. It is shown in Kalton [1977a] and [1978e] that certain Orlicz spaces L_φ have separating dual, but the kernel of any compact operator is weakly closed.

To conclude the section we construct a class of spaces which have 'many' compact operators. We define a quasi-Banach space X to be <u>pseudo- reflexive</u> if there is a Hausdorff vector topology α on X so that

(1) The unit ball B of X is α-relatively compact;
(2) Every closed linear subspace is also α-closed.

If X is pseudo-reflexive, then it has enough compact operators to separate points from closed subspaces. Precisely, if N is a closed subspace of X and $x \notin N$ then there is a compact operator $K : X \to Y$ so that $K(N) = 0$ and $Kx \neq 0$. We leave this exercise to the reader.

EXAMPLE. A non-locally convex pseudo-reflexive space.

Fix p, $0 < p < 1$, and let $E_n = \ell_p^n$, the n-dimensional ℓ_p-space. Let $X = \ell_2(E_n)$, the space of sequences (x_n) where $x_n \in E_n$ and

$$\|(x_n)\| = (\sum_{n=1}^{\infty} \|x_n\|^2)^{1/2} < \infty ;$$

$\ell_2(E_n)$ is a non-locally convex p-Banach space (see Section 4.4).

Now let α be the topology on X of convergence in each co-ordinate and consider $\hat{\alpha}$. It is not difficult to show that B is α-compact where B is the unit ball of X.

Let E be a closed linear subspace. We check that $E \cap B$ is α-closed, so that by Theorem 7.17, E is $\hat{\alpha}$-closed. Suppose on the contrary that $u_n \in E \cap B$ and $u_n \to u(\alpha)$ where $u \notin E$. Then $u - u_n \to 0(\alpha)$, but $\|u-u_n\| \geq d(u,E) > 0$ for all n. By a standard gliding hump argument we can find a subsequence $(u-u_{n_k})$ which is basic and equivalent to a sequence (v_k) with disjoint supports on $\mathbf{N}$. However (v_k) is equivalent to the usual basis of ℓ_2. Thus $\sum (1/k)(u-u_{n_k})$ converges. However, for all $n \in \mathbf{N}$,

$$\| \sum_{k=1}^{n} (1/k)(u-u_{n_k})\| \geq (\sum_{k=1}^{n} (1/k))d(u,E).$$

This contradiction shows that E is $\hat{\alpha}$-closed and so X is pseudo-reflexive. □

EXAMPLE. A pseudo-reflexive space with trivial dual.

The space X of the preceding example has a PCWD-subspace N (Chapter 4). Then X/N is also pseudo-reflexive (use the quotient $\hat{\alpha}$-topology) and has trivial dual. Of course this is yet another space with trivial dual and plenty of compact operators.

5. Operators on L_p

We have seen that there are no non-zero compact or even strictly singular endomorphisms of L_p when $0 < p < 1$. The results of Section 4 suggest we should examine whether L_p can be the domain of a strictly singular operator. It turns out that the answer to this question is no (Kalton, 1977a).

THEOREM 7.20. Let Y be a quasi-Banach space and $T : L_p \to Y$ be a non-zero operator. Then there is a subspace H of L_p with $H \approx \ell_2$ so that $T|_H$ is an isomorphism.

REMARK 1. This theorem is true for Y any F-space with some rewording of the proof.

REMARK 2. L_p $(0 < p < 1)$ does not admit compact or even strictly singular operators. Compare Theorem 2.2.

Proof. Let us suppose on the contrary that T is not an isomorphism on any closed subspace $H \approx \ell_2$. Denote by 1 the constantly one function on $(0,1)$. We assume Y has an r-subadditive quasi-norm for some $r > 0$.

Let $(\eta_j : j = 1,2,\ldots)$ be a sequence of independent random variables with normal distribution, mean zero and variance one. Then $H = [\eta_j]$ is isomorphic to ℓ_2. If $\varphi \in H$ and $\|\varphi\| = \|\eta_1\|$ then φ has the same distribution as η_1.

For $S \in L(L_p)$ with $\|S\| \leq 1$, and $\epsilon > 0$,

$$\|TS1\|^r \leq \|TS(1+\epsilon\varphi)\|^r + \epsilon^r\|TS\varphi\|^r$$

for any $\varphi \in L_p$.

Thus

$$\|TS1\|^r \leq \|T\|^r\|1+\epsilon\varphi\|^r + \epsilon^r\|TS\varphi\|^r.$$

Now $TS|_H$ cannot be an isomorphism - for otherwise T is an isomorphism on $S(H) \approx H \approx \ell_2$. Hence we can choose $\varphi \in H$ with $\|\varphi\| = \|\eta_1\|$ and $\|TS\varphi\|$ as small as we please. Thus

$$\|TS1\| \leq \|T\| \; \|1+\epsilon\eta_1\|.$$

Now allow S to vary over all operators with $\|S\| \leq 1$. From the transitivity properties of L_p we have that

$$\sup_{\|S\|\leq 1} \|TS1\| = \|T\|.$$

Hence

$$\|T\| \leq \|T\| \; \|1+\epsilon\eta_1\|$$

or

$$\|1+\epsilon\eta_1\| \geq 1.$$

This last statement is then true for every $\epsilon > 0$. Now

$$\|1+\epsilon\eta_1\|^p = (1/\sqrt{2\pi}) \int_{-\infty}^{\infty} |1+\epsilon x|^p e^{-x^2/2} dx$$

$$= (1/\sqrt{2\pi}) \int_{-\infty}^{\infty} |1-\epsilon x|^p e^{-x^2/2} dx .$$

Hence

$$\int_{-\infty}^{\infty} (|1+\epsilon x|^p + |1-\epsilon x|^p - 2)/(2\epsilon^2) \; e^{-1/2\; x^2} dx \geq 0.$$

Now for some constant M,

$$||1+u|^p + |1-u|^p - 2| \leq Mu^2$$

and hence

$$|(|1+\epsilon x|^p + |1-\epsilon x|^p - 2)/(2\epsilon^2)| \leq (1/2)Mx^2.$$

But

$$(1/\sqrt{2\pi}) \int_{-\infty}^{\infty} x^2 e^{-x^2/2} dx = 1,$$

and so the dominated convergence theorem can be applied. Letting $\epsilon \to 0$,

$$(|1+\epsilon x|^p + |1-\epsilon x|^p - 2)/(2\epsilon^2) \to (p(p-1)/2)x^2$$

and hence

$$(p(p-1))/2 \int_{-\infty}^{\infty} x^2 e^{-x^2/2} dx \geq 0.$$

This is a contradiction as $p-1 < 0$. □

Theorem 7.20 was actually proved for a wider class of Orlicz spaces, and may be extended to certain symmetric function spaces. To prove that all compact operators on L_p are zero is actually much easier - a very simple proof for symmetric spaces was given in Kalton [1978d]. We sketch the details. Let $K : L_p \to X$ be a compact operator. Since L_p is transitive we need only show that $K(1) = 0$. For each $\epsilon > 0$ let (A_n) be a sequence of independent sets of measure ϵ; let χ_n be their characteristic functions. By passing to a subsequence we may suppose that $(K\chi_n)$ converges to a limit $u \in X$ rather fast, i.e.

$$\|K\chi_n - u\| \leq 2^{-n}.$$

Now $\epsilon^{-1/2}(1-\epsilon)^{-1/2}(\chi_n - \epsilon)$ is an orthonormal sequence in L_2 and so $\sum (1/n)(\chi_n - \epsilon)$ converges in L_2 and hence in L_p. Thus

$$\sum (1/n)(K\chi_n - \epsilon K(1))$$

converges, so that

$$\sum (1/n)(u - \epsilon K(1))$$

converges, i.e. $u = \epsilon K(1)$. Thus

$$\begin{aligned} \|K(1)\| &\leq \epsilon^{-1}\|u\| \\ &\leq \epsilon^{-1}\|K\|\,\|\chi_n\| \qquad \text{(each } n) \\ &= \epsilon^{-1}\|K\|\epsilon^{1/p}. \end{aligned}$$

As $\epsilon > 0$ is arbitrary $K(1) = 0$.

In the next chapter we shall discuss whether Theorem 7.20 can be improved by replacing ℓ_2 by any ℓ_q $(p < q \leq 2)$ or even L_q. However in the remaining part of this chapter we present a theorem extending Theorem 7.20 partially to a larger class of domain spaces. We return to our initial approach in this chapter and consider transitivity hypotheses.

DEFINITION. A quasi-Banach space X is <u>boundedly transitive</u> if there is a constant $M < \infty$ such that if $x,y \in X$ with $\|x\| = \|y\| = 1$ then there exists $T \in L(X)$ with $Tx = y$ and $\|T\| \leq M$.

THEOREM 7.21. Let X be a non-locally convex boundedly transitive quasi-Banach space. Let Y be any quasi-Banach space and let $K : X \to Y$ be a compact operator. Then $K = 0$.

<u>Proof</u>. Let us recall that the sequence $a_n = a_n(X)$ is defined by

$$a_n = \sup (\|x_1+\ldots+x_n\| : \|x_i\| \leq 1, \quad x_i \in X).$$

Now X is not locally convex so that $n^{-1}a_n \to \infty$. We recall that $a_{mn} \leq a_m a_n$ $(m,n \in \mathbf{N})$ and (a_n) is monotone increasing.

We claim that X has property (P)

(P): if $u \in X$ with $\|u\| = 1$ then there exist $v_1,\ldots,v_n$ with $\|v_i\| \leq 2Ma_n^{-1}$ and $u = v_1+\ldots v_n$.

To establish (P) simply choose any $x_1,\ldots,x_n$ with $\|x_i\| \leq 1$ and

$$\|x_1+\ldots+x_n\| \geq (1/2)a_n.$$

Choose $S \in L(X)$ with $\|S\| \leq M\|x_1+\ldots+x_n\|^{-1}$ so that $S(x_1+\ldots+x_n) = u$ and let $v_i = Sx_i$.

If we apply (P) twice to fixed $u \in X$ we can find for $m,n \in \mathbb{N}$, $(w_{ij} : 1 \leq i \leq m,\ 1 \leq j \leq n)$ so that $\|w_{ij}\| \leq 4M^2 a_m^{-1} a_n^{-1}$ and

$$u = \sum_{i=1}^{m} \sum_{j=1}^{n} w_{ij}.$$

Thus

$$\begin{aligned} 1 &= \|u\| \\ &\leq a_{mn} \max\|w_{ij}\| \\ &\leq 4M^2 a_{mn}/a_m a_n \end{aligned}$$

or

$$a_m a_n \leq 4M^2 a_{mn}.$$

Now by induction we have

$$a_{n^k} \leq a_n^k \leq (4M^2)^{k-1} a_{n^k} \qquad n,k \in \mathbb{N}.$$

or

$$(a_{n^k})^{1/k} \leq a_n \leq 4M^2 (a_{n^k})^{1/k}.$$

Now the properties of the sequence (a_n) imply easily that

$$\lim_{n\to\infty} (\log a_n)/(\log n) = \alpha$$

exists. (Simply observe that $\log a_{2^k}$ is subadditive and hence $\lim_{k\to\infty} k^{-1}\log a_{2^k}$ exists. Then use the fact that (a_n) is monotone.)

Since $n \leq a_n$, we have $\alpha \geq 1$. Let $\alpha = 1/p$ where $0 < p \leq 1$. Then

$$\lim_{k\to\infty} (a_{n^k})^{1/k} = n^{1/p}.$$

Hence

$$n^{1/p} \leq a_n \leq 4M^2 n^{1/p} \qquad n \in \mathbf{N}.$$

In particular $p < 1$ as $a_n/n \to \infty$.

Now suppose $K : X \to Y$ is a compact operator. For $n \in \mathbf{N}$ let $\mathcal{B}_n$ be the subalgebra of the Borel sets of $(0,1)$ generated by the dyadic intervals $((k-1)2^{-n},k2^{-n})$ $(1 \leq k \leq 2^n)$. Let $L_p(\mathcal{B}_n)$ be the subspace of L_p of all $f \in L_p$ measurable with respect to $\mathcal{B}_n$. Let $V = K(B)$ where B is the unit ball of X.

Fix any $u \in X$, with $\|u\| = 1$. We shall show that $Ku = 0$.

For $n \in \mathbf{N}$, let $N = 2^n$ and write

$$u = x_1 + \ldots + x_N$$

where $\|x_i\| \leq 2Ma_N^{-1} \leq 2MN^{-1/p}$.

Define $T_n : L_p(\mathcal{B}_n) \to Y$ to be linear such that

$$T_n \chi_{k,n} = Kx_k \qquad 1 \leq k \leq 2^n$$

where $\chi_{k,n}$ is the characteristic function of $((k-1)2^{-n}, k2^{-n})$.

Suppose $f \in L_p(\mathcal{B}_n)$ with $\|f\| \leq 1$. We shall show that $T_n f \in 16M^4 V$. To do this it suffices to consider the dense set of such f with $|f|^p$ rational-valued, say

$$f = \sum_{k=1}^{N} \epsilon_k (m_k/m)^{1/p} \chi_{k,n}$$

where $\epsilon_k = \pm 1$, $m_k \in \mathbf{N}$ and $\sum m_k \leq mN$.

By property (P) we can write

$$x_k = \sum_{i=1}^{m_k} y_{ki}$$

where $\|y_{ki}\| \leq 2Ma_{m_k}^{-1} \|x_k\| \leq 4M^2 m_k^{-1/p} N^{-1/p}$. Then

$$\|\epsilon_k (m_k m^{-1})^{1/p} y_{ki}\| \leq 4M^2 m^{-1/p} N^{-1/p}$$

and

$$\left\| \sum_{k=1}^{N} \sum_{i=1}^{m_k} \epsilon_k (m_k m^{-1})^{1/p} y_{ki} \right\| \leq a_{mN} (4M^2 m^{-1/p} N^{-1/p})$$
$$\leq 16M^4.$$

Thus

$$T_n f = K\left(\sum_{k=1}^{N} \sum_{i=1}^{m_k} \epsilon_k (m_k m^{-1})^{1/p} y_{ki} \right)$$
$$\in 16M^4 V.$$

Now let Z be the union of the spaces $L_p(\mathcal{B}_n)$ for $n \in \mathbf{N}$. Define $T : Z \to Y$ by

$$Tf = \lim_{n \in \mathcal{U}} T_n f$$

where $\mathcal{U}$ is some free ultrafilter on **N**. Clearly T is well defined on Z and $Tf \in 16M^4V$ if $\|f\| < 1$. Thus T extends to a compact operator on $L_p(0,1)$, and so $T = 0$. However

$$\begin{aligned} T1 &= \lim_{\mathcal{U}} Ku \\ &= Ku \\ &= 0. \end{aligned}$$

Hence $Ku = 0$ and so $K = 0$.

6. A rigid space

In this last section we construct a quasi-Banach space which fails to be transitive in the strongest possible way. In fact we make a _rigid space_, i.e. a space X such that $\mathcal{L}(X) = \{\lambda I : \lambda \in \mathbf{K}\}$ where $\mathbf{K}$ is the scalar field.

Before doing this, let us note a relationship with some problems raised in Chapter 4. Recall that an infinite-dimensional quasi-Banach space X is _atomic_ if every proper closed subspace of X is finite-dimensional.

THEOREM 7.22. Any atomic complex quasi-Banach space X is rigid.

Proof. If $T \in \mathcal{L}(X)$ and $T \neq 0$ then T has closed range since X is quotient-minimal. Clearly T cannot have finite-rank since $X^* = \{0\}$. Hence T is surjective and so $T \in G_2(X;X)$. Hence $G_2(X;X) = \mathcal{L}(X) \setminus \{0\}$. If $\dim \mathcal{L}(X) > 1$, then G_2 is connected and by Proposition 7.9, $G_1 = G_2$. Hence any maximal abelian subalgebra of $\mathcal{L}(X)$ is a field and Theorem 7.2 quickly yields that $\mathcal{L}(X) = \{\lambda I : \lambda \in \mathbf{C}\}$. □

In the real case the same argument shows that $L(X)$ is isomorphic to $\mathbb{R}$, $\mathbb{C}$ or the quaternions $\mathbb{Q}$.

In order to construct a rigid F-space we introduce a device which is often useful in constructing F-spaces. Suppose V is a vector space, $C > 0$, $\{(V_\alpha, \|\cdot\|_\alpha) : \alpha \in I\}$ is a collection such that each V_α is a subspace of V and each $\|\cdot\|_\alpha$ is a quasi-F-norm on V_α such that for every $x,y \in V_\alpha$

$$\|x+y\|_\alpha \leq C(\|x\|_\alpha + \|y\|_\alpha),$$

and

$$\text{span}\,\{V_\alpha : \alpha \in I\} = V.$$

We define $\|\cdot\| = \inf\{\|\cdot\|_\alpha : \alpha \in I\}$ on V by

$$\|x\| = \inf\{\|x_{\alpha_1}\|_{\alpha_1} + \ldots + \|x_{\alpha_n}\|_{\alpha_n} : x_{\alpha_i} \in V_{\alpha_i} \text{ and } \sum_{i=1}^{n} x_{\alpha_i} = x\}.$$

It is easily verified that if $\|x\| > 0$ for $x \neq 0$, then $\|\cdot\|$ is a quasi-F-norm on V and

$$\|x+y\| \leq C(\|x\|+\|y\|) \quad \text{for every } x,y \in V.$$

If, in addition, each $\|\cdot\|_\alpha$ is a quasi-norm, then $\|\cdot\|$ is a quasi-norm.

We now turn to the actual construction of a rigid space.

Let V be the set of all sequences that are eventually zero, let $\{e_n : n = 1,2,\ldots\}$ be the coordinate vectors and let

$$A = \{e_1+e_n : n = 2,3,\ldots\}$$
$$\cup \{e_1-e_n : n = 2,3,\ldots\} \cup \{e_1\}.$$

Also let (a_n) be a sequence in A so that for every $a \in A$, $a_n = a$ for infinitely many n. We now inductively select a sequence $(F_n, V_n, \|\cdot\|_n, \epsilon_n)$ where each F_n is a finite subset of V, $V_n = \text{span } F_n$, $\|\cdot\|_n$ is a quasi-norm on V_n, and $\epsilon_n \downarrow 0$. Also let $|||\cdot|||_n = \inf\{\|\cdot\|_1, \ldots, \|\cdot\|_n\}$ on $V_1 + \ldots + V_n$. We make this selection so that the following conditions hold:

(X1) $a_n \in V_n$ and $(V_1 + \ldots + V_{n-1}) \cap V_n \subset \mathbb{R}a_n$

(X2) $|||\cdot|||_m > (1/2)|||.|||_n$ on $V_1 + \ldots + V_n$ if $m > n$

(X3) If $B_r^n = \{x \in V_1 + \ldots + V_n : |||x|||_n < r\}$

for $r > 0$, then $\text{co } B^n_{\epsilon_{n+1}} \subset B^n_{\epsilon_n}$

(X4) If $F_n = \{x_1, \ldots, x_m\}$, then $\|x_i\|_n < \epsilon_n$ and

$$1/n \sum_{i=1}^{n} x_i = a_n.$$

(X5) If $M_n = \max\{|F_1|, \ldots, |F_{n-1}|\}$ for $n \geq 2$, $r \leq M_n$

and $y_1, \ldots, y_r \in V_n$, then

$$\|y_1+\ldots+y_r\|_n \leq 2(\|y_1\|_n + \ldots + \|y_r\|_n).$$

First choose $(F_1, V_1, \|\cdot\|_1, \epsilon_1)$ to satisfy (X1) and (X4) and so that if $x, y \in V_1$, then

$$\|x+y\|_1 \leq 2(\|x\|_1+\|y\|_1)$$

(each $\|\cdot\|_n$ will be chosen to satisfy this.)

Suppose $(F_i, V_i, \|\cdot\|_i, \epsilon_i)$ have been selected for $1 \leq i \leq n$. Since $V_1 + \ldots + V_n$ is finite dimensional, there exists ϵ_{n+1} such that $0 < \epsilon_{n+1} < \min\{\epsilon_n, 1/n\}$ so that

$$\text{co } B_{\epsilon_{n+1}}{}^n \subset B_{\epsilon_n}{}^n.$$

Let $1 \leq k \leq n$ and let $S_k = \{x \in V_1+\ldots+V_k : |||x|||_k = 1\}$. Define $|\cdot|_\beta$ on $\mathbb{R}a_{n+1}$ by $|\lambda a_{n+1}| = \beta|\lambda|$ for $\beta > 0$. Now for $\beta \uparrow$,

$$\inf\{|||\cdot|||_n, |\cdot|_\beta\} \uparrow |||\cdot|||_n$$

on S_k; and since S_k is compact, the convergence is uniform by Dini's Theorem. By our induction hypothesis $|||\cdot|||_n > (1/2)|||\cdot|||_k$, $1 \leq k \leq n$. Thus for β chosen suitably large

$$\inf\{|||\cdot|||_n, |\cdot|_\beta\} > (1/2)|||\cdot|||_k$$

$$\text{on } V_1 + \ldots + V_k \text{ for } 1 \leq k \leq n.$$

Now choose $0 < p < 1$ with p suitably close to 1 so that if $r \leq M_{n+1}$ and $y_1, \ldots, y_r \in \ell_p$ then

$$\|y_1+\ldots+y_r\|_p \leq 2(\|y_1\|_p+\ldots+\|y_r\|_p)$$

where $\|\cdot\|_p$ denotes the standard quasi-norm on ℓ_p. Also select an integer m so that

$$\|(\epsilon_{n+1}/2m)(e_1+\ldots+e_m)\|_p = (\epsilon_{n+1}/2)m^{1/p-1} \geq \beta.$$

One can find V_{n+1} in V so that $a_{n+1} \in V_{n+1}$, $(V_1+\ldots+V_n) \cap V_{n+1} \subset Ra_{n+1}$ and $\dim V_{n+1} = m$. Also one can define a linear operator

$$T : \ell_p^m \xrightarrow[\text{onto}]{1-1} V_{n+1}$$

so that $T((\epsilon_{n+1}/2m)(e_1+\ldots+e_m)) = a_{n+1}$. Now let $x_i = T((\epsilon_{n+1}/2)e_i)$ for $1 \leq i \leq m$, $F_{n+1} = \{x_1,\ldots,x_m\}$ and for $x \in \ell_p^m$, $\|T(x)\|_{n+1} = \|x\|_p$. Notice that $\|\cdot\|_{n+1} \geq |\cdot|_\beta$ on Ra_{n+1}.

Now suppose $x \in V_1 + \ldots + V_k$ for $1 \leq k \leq n$. If $x = x_1 + x_2$ where $x_1 \in V_1 + \ldots + V_k$ and $x_2 \in V_{n+1}$, then $x_2 = x - x_1 \in (V_1+\ldots+V_k) \cap V_{n+1}$ so that $x_2 = \lambda a_{n+1}$ for some constant λ. Thus

$$|||x_1|||_n + \|x_2\|_{n+1} \geq |||x_1|||_n + |\lambda a_{n+1}|_\beta > (1/2)|||x|||_k$$

since $\inf\{|||\cdot|||_n, |\cdot|_\beta\} > (1/2)|||\cdot|||_k$ on $V_1+\ldots+V_k$. Hence (X2) holds for n+1. Notice that conditions (X1), (X3), (X4), and (X5) also hold for n+1.

Finally, let

$$\|\cdot\| = \inf\{\|\cdot\|_n : n = 1,2,\ldots\}$$
$$= \inf\{|||\cdot|||_n : n = 1,2,\ldots\}.$$

Observe that if $x,y \in V$, then $\|x+y\| \leq 2(\|x\|+\|y\|)$ and by condition (X2), $\|x\| = 0$ only if $x = 0$. $(X,\|\cdot\|)$ will now denote the completion of $(V,(\|\cdot\|)$. We claim that X is rigid.

THEOREM 7.23. $(X,\|\cdot\|)$ is rigid.

<u>Proof</u>. Let $T \in L(X)$. We shall prove that if $a \in A$, then there exists a constant c so that $T(a) = ca$. This will prove the result, for suppose $T(e_1) = \lambda e_1$ and for $n \geq 2$, $T(e_1+e_n) = \lambda_n(e_1+e_n)$ and $T(e_1-e_n) = \lambda_n'(e_1-e_n)$. Then

$$2\lambda e_1 = T(2e_1) = T(e_1+e_n) + T(e_1-e_n)$$
$$= \lambda_n(e_1+e_n) + \lambda_n'(e_1-e_n) = (\lambda_n+\lambda_n')e_1 + (\lambda_n-\lambda_n')e_n.$$

Hence $\lambda_n = \lambda_n' = \lambda$. Thus for every $a \in A$, $T(a) = \lambda a$. Since X is the closed span of A, $T = \lambda I$.

We may assume that

$$\|T\| < 1/2.$$

Let n be an integer with $n \geq 2$. Since $\bar{V} = X$ there exists $S : V_n \to V$ such that S is linear, if $F_n = \{x_1,\ldots,x_m\}$ then

$$\|S(x_i)\| < \epsilon_n$$

and

$$\|S(a_n)-T(a_n)\| < \epsilon_n.$$

Recall that $a_n = 1/m \sum_{i=1}^{m} x_i$. Since $\|S(x_i)\| < \epsilon_n$ and $S(x_i) \in V$, there exist $x_{i1},\dots,x_{ik}$ with $x_{ij} \in V_j$ such that

$$S(x_i) = x_{i1} + \dots + x_{ik} \quad \text{and} \quad \sum_{j=1}^{k} \|x_{ij}\|_j < \epsilon_n.$$

(We may assume $k > n$ by taking $x_{ij} = 0$ for some j's if necessary.) Thus

$$S(a_n) = 1/m \sum_{i=1}^{m} (x_{i1}+\dots+x_{ik}).$$

Since $|||x_{i1}+\dots+x_{i(n-1)}|||_{n-1} < \epsilon_n$,

$$\|1/m \sum_{i=1}^{m} (x_{i1}+\dots+x_{i(n-1)})\|$$

$$\leq |||1/m \sum_{i=1}^{m} (x_{i1}+\dots+x_{i(n-1)})|||_{n-1}$$

$$< \epsilon_{n-1}$$

by condition (X3). If $j > n$, by condition (X5),

$$\|1/m \sum_{i=1}^{m} x_{ij}\|_j \leq 2/m \sum_{i=1}^{m} \|x_{ij}\|_j.$$

Hence,

$$\|1/m \sum_{i=1}^{m} (x_{i(n+1)}+\dots+x_{ik})\|$$

$$\leq |||1/m \sum_{i=1}^{m} (x_{i(n+1)} + \dots + x_{ik})|||_k$$

$$\leq 1/m(\| \sum_{i=1}^{m} x_{i(n+1)}\|_{n+1} + \dots + \| \sum_{i=1}^{m} x_{ik}\|_k)$$

$$\leq 2/m\left(\sum_{i=1}^{m} \|x_{i(n+1)}\|_{n+1} + \ldots + \sum_{i=1}^{m} \|x_{ik}\|_k\right)$$

$$< 2\epsilon_n.$$

Let $y_n = 1/m \sum_{i=1}^{m} x_{in} \in V_n$. Then

$$\|S(a_n)-y_n)\| \leq$$

$$\|1/m \sum_{i=1}^{m} (x_{i1}+\ldots+x_{i(n-1)}) + 1/m \sum_{i=1}^{m} (x_{i(n+1)}+\ldots+x_{1k})\|$$

$$< 2(\epsilon_{n-1}+2\epsilon_n) = 2\epsilon_{n-1} + 4\epsilon_n.$$

Thus

$$\|T(a_n)-y_n\| \leq 2(\|T(a_n)-S(a_n)\| + \|S(a_n)-y_n\|)$$

$$< 2(\epsilon_n+2\epsilon_{n-1}+4\epsilon_n) < 14\epsilon_{n-1}.$$

Now let $a \in A$ and suppose $a_m = a_n = a$ where $2 \leq m < n$. Then

$$|||y_m-y_n|||_n \leq 4(||y_m-T(a)||_n + ||y_n-T(a)||_n$$

$$< 4(14\epsilon_{m-1}+14\epsilon_{n-1}) < 112\epsilon_{m-1}.$$

Thus $y_m-y_n = x_1 + \ldots + x_n$ with $x_i \in V_i$ and $\sum_{k=1}^{n} \|x_i\|_i < 112\epsilon_{m-1}$. Since $m < n$, $y_m -(x_1+\ldots+x_{n-1}) = y_n+x_n \in (V_1+\ldots+V_{n-1}) \cap V_n = Ra$. Hence there exists a constant λ so that $y_m-(x_1+\ldots+x_{n-1}) = \lambda a$ and therefore $\|y_m-\lambda a\| < \|x_1\|_1 + \ldots + \|x_{n-1}\|_{n-1} < 112\epsilon_{m-1}$. Consequently

$$\|T(a)-\lambda a\| < 2(\|T(a)-y_m\|+\|y_m-\lambda a\|)$$

$$< 2(14\epsilon_{m-1}+112\epsilon_{m-1}) = 252\epsilon_{m-1}.$$

Since $a_m = a$ for infinitely many m, $T(a) \in Ra$. □

REMARKS. A refinement of this construction yields a rigid subspace of L_p for $0 \leq p < 1$ (see Kalton-Roberts (1981)). In fact there exists a rigid subspace X of L_p such that every non-trivial quotient of X is also rigid. Also there is a continuum of non-isomorphic rigid subspaces of L_p when $0 \leq p < 1$.

CHAPTER 8
OPERATORS BETWEEN L_p SPACES, $0 < p < 1$

1. Introduction

In this chapter we present some representation theorems for operators from L_p to $L_p(\mu)$, $0 < p < 1$. The theorems have some important consequences; for example, we will show that a non-zero operator from L_p to $L_p(\mu)$, $0 < p < 1$, is an isomorphism when restricted to $L_p(A)$, for some set A of positive measure.

From Theorem 7.12 of the previous chapter, any non-zero endomorphism of L_p, $0 < p < 1$, is an isomorphism on some infinite-dimensional subspace - and by Theorem 7.20 of the previous chapter, the subspace can be taken to be ℓ_2. We are now asserting considerably more. Our second assertion above trivially implies that a non-zero operator from L_p into $L_p(\mu)$ preserves a copy of ℓ_2, since ℓ_2 embeds isomorphically into $L_p(A)$. Of course it also implies Pallaschke's original result that for $0 < p < 1$, L_p admits no non-trivial compact endomorphisms.

Pallaschke's results on the endomorphisms of L_p, $0 < p < 1$, appeared in 1973. A further step was taken by Berg-Peck-Porta [1973], who studied projections on L_0. Kwapien [1973] characterized completely the operators from L_0 to $L_0(\mu)$.

Then Kalton [1978a] characterized completely the operators from L_p to $L_p(\mu)$, $0 < p < 1$, and derived a number of results on the structure of L_p, $0 \leq p < 1$, including the above-mentioned one, as corollaries.

2. Operators between L_0-Spaces

Our approach to the study of operators between L_0 and $L_0(\mu)$ is a somewhat combinatorial version of Kwapien's approach. As before, (Ω,Σ,μ) is a finite measure space; throughout this section we assume that Σ is μ-complete. Also, λ denotes Haar measure on Δ, the Cantor set, and M denotes the λ-measurable subsets of Δ.

We begin with some basic facts about operators from L_0 to $L_0(\mu)$. For a function f let supp f be the set where f is non-zero.

LEMMA 8.1. Suppose T is an operator from L_0 to $L_0(\mu)$. Then for every $\epsilon > 0$ there is a $\delta > 0$ so that if $\lambda(\text{supp } f) < \delta$ then $\mu(\text{supp } Tf) < \epsilon$.

Proof. Given $\epsilon > 0$, choose $\delta > 0$ so that $\|f\|_0 < \delta \Longrightarrow \|Tf\|_0 < \epsilon/2$. Now suppose $\lambda(\text{supp } f) < \delta$. Then for any $n \in \mathbf{N}$ $\|nf\|_0 < \delta$, so

$$\|T(nf)\|_0 = \int_\Omega (|nTf|)/1+|nTf|) \, d\mu < \epsilon/2.$$

The function in the integrand converges pointwise to $1_{\text{supp } Tf}$; using the dominated convergence theorem we obtain

$$\mu(\text{supp } Tf) \leq \epsilon/2 < \epsilon. \qquad \square$$

Next we examine a very simple kind of operator. Suppose $\sigma : \Omega \to \Delta$ is Σ-$\mathcal{M}$ measurable. Define $T : L_0 \to L_0(\mu)$ by

$$Tf = f \circ \sigma.$$

In order for T to be well-defined, it is necessary that whenever $f = 0$ a.e. then $Tf = 0$ a.e. In particular, suppose $\lambda(A) = 0$. Then

$$T(1_A) = 1_{\sigma^{-1}(A)} = 0, \quad \mu \text{ a.e.}$$

so

$$\mu(\sigma^{-1}(A)) = 0.$$

That is, σ is non-singular according to the following:

DEFINITION. A Σ-$\mathcal{M}$ measurable map from Ω to Δ is <u>non-singular</u> if whenever $A \in \mathcal{M}$ and $\lambda(A) = 0$, then $\mu(\sigma^{-1}(A)) = 0$.

Note that once σ is non-singular, the operator T defined above is automatically continuous. Indeed, suppose $\epsilon > 0$. Since σ is non-singular, there is $\delta > 0$ such that $\lambda(A) < \delta$ implies $\mu(\sigma^{-1}(A)) < \epsilon$. Now suppose $\lambda\{|f| > \delta\} < \delta$. Then $\mu\{|f \circ \sigma| > 0\} < \epsilon$, i.e. $\mu\{|Tf| > \delta\} < \epsilon$, so T is continuous.

We can construct a slightly more general class of operators as follows: Suppose that $\sigma : \Omega \to \Delta$ is non-singular and that $g : \Omega \to \mathbf{R}$ is measurable. Define $T' : L_0 \to L_0(\mu)$ by

$$(T'f)(x) = g(x)f(\sigma(x)).$$

The operator T is continuous, since $L_0(\mu)$ is a topological algebra.

At this point we can give an informal statement of Kwapien's theorem: every linear operator from L_0 to $L_0(\mu)$ is an infinite sum of operators of the form T' above.

A further fact we will need is:

LEMMA 8.2. Let $(f_i)_{i=1}^n$ be a finite sequence in $L_0(\mu)$. Let $A_i = \operatorname{supp} f_i$, $1 \leq i \leq n$. Then there are scalars $(c_i)_{i=1}^n$ such that if $f = \sum_{i=1}^n c_i f_i$, $\mu(\bigcup_{i=1}^n A_i \sim \operatorname{supp} f) = 0$.

<u>Proof</u>. It suffices to prove this when $n = 2$. The general case follows by induction.

For each real number r, let

$$S_r = A_2 \cap \{x : f_1(x) + rf_2(x) = 0\}.$$

Note that if $r \neq s$, then $S_r \cap S_s = \phi$. The sets $\{S_r : r \in \mathbf{R}\}$ are an uncountable collection of pairwise disjoint members of Σ; since μ is finite, at least one of these sets has measure zero, say S_{r_0}. Then if $x \in (A_1 \cup A_2 \sim \operatorname{supp}(f_1+r_0f_2))$, it follows easily that x is in the null set S_{r_0}. □

Next, we need to mention some properties of the measure algebras (Δ, M, λ) and (Ω, Σ, μ) and of non-singular maps from Ω to Δ.

NOTATION. For $A \in \mathcal{M}$, denote by $\mathcal{M}(A)$ the restriction of the measure algebra $\mathcal{M}$ to A. Similarly the restriction of Σ to a set B in Σ will be denoted by $\Sigma(B)$.

(i) Recall the standard definition of distance between two sets in $\mathcal{M}$ (respectively, Σ):

$$d_1(C,D) = \lambda(C\Delta D)$$

$$d_2(C,D) = \mu(C\Delta D).$$

Note that d_1 and d_2 are complete metrics.

(ii) L_0 is a complete lattice. In fact, if $A \subset L_0$ there is an extended real valued measurable function f such that

(a) $A \leqslant f$, i.e. if $h \in A$, then $h \leqslant f$ a.e.

(b) $A \leqslant g$ for an extended real valued measurable function g implies that $f \leq g$ a.e.

f is uniquely determined by (a) and (b) and is called the least upper bound of A. We denote f by sup A. Furthermore, there is a countable subcollection $B \subset A$ such that sup B = sup A. The proof of this is elementary and left as as an exercise. The analogous result for subcollections of Σ is a corollary.

(iii) In any measure space (Ω,Σ,μ) we shall not distinguish between Σ and its corresponding measure algebra.

DEFINITION. Suppose $(\Omega_1,\Sigma_1,\mu_1)$ and $(\Omega_2,\Sigma_2,\mu_2)$ are measure spaces. A <u>continuous Boolean homomorphism</u> from Σ_1 to Σ_2 is a map $\Phi : \Sigma_1 \to \Sigma_2$ satisfying

(a) $\Phi(\phi) = \phi$,

(b) $\Phi(A \cup B) = \Phi(A) \cup \Phi(B)$,

$\Phi(A \cap B) = \Phi(A) \cap \Phi(B)$, $A, B \in \Sigma_1$

(c) $\mu_1(A_n) \to 0 \Longrightarrow \mu_2(\Phi(A_n)) \to 0$.

(These statements hold in the measure algebras; alternatively, they are valid up to sets of measure zero. Similar comments apply below.)

DEFINITION. $(\Omega_1, \mu_1), (\Omega_2, \mu_2)$ is a _regular pair_ of measure spaces if the following condition holds for every $E \in \Sigma_1$ and $F \in \Sigma_2$: whenever $\Phi : \Sigma_1(E) \to \Sigma_2(F)$ is a continuous Boolean homomorphism, there is a non-singular $\Sigma_2(F) - \Sigma_1(E)$-measurable map σ from F to E such that

$$\Phi(A) = \sigma^{-1}(A) \quad \text{for all } A \text{ in } \Sigma_1(E).$$

It follows from (i) above and our assumptions on (Ω, Σ, μ) that (Δ, M, λ), (Ω, Σ, μ) _is a regular pair of measure spaces_. (See Royden [1963, Theorem 14.10].)

There is one more technical lemma which we need. It proves a combinatorial version of Lemma 2 of Kwapien (1973).

LEMMA 8.3. Suppose $A_1, \ldots, A_n$, B are in Σ and that $M \in \mathbf{N}$ is such that

$$\sum_{i=1}^{n} 1_{A_i} \geq M 1_B.$$

Then for $1 \leq r \leq n$, there is a subfamily $A_{i_1}, \ldots, A_{i_r}$ of (A_i) such that

$$\mu\left(\bigcup_{j=1}^{r} A_{i_j}\right) \geq (1 - (1 - (r/n))^M)\mu(B).$$

<u>Proof</u>. Let $t \in B$. Then

$$\sum_{|E|=r} 1_{\bigcup_{i\in E} A_i}(t) \geq \binom{n}{r} - \binom{n-M}{r}.$$

To see this, note that there is $F \subset \{1,\ldots,n\}$ with $|F| = M$ such that $t \in \bigcap_{i\in F} A_i$. Then if $|E| = r$, $t \in \bigcup_{i\in E} A_i$ only if $E \cap F = \phi$.

There are $\binom{n-M}{r}$ r-sets E failing to intersect F, which proves the claim.

Thus the average measure of a union of r-many sets A_i is

$$\binom{n}{r}^{-1} \sum_{|E|=r} \mu(\bigcup_{i\in E} A_i) \geq \binom{n}{r}^{-1} \int_{E_0} \Big(\sum_{|E|=r} 1_{\bigcup_{i\in E} A_i}\Big) d\mu$$

$$\geq \binom{n}{r}^{-1} \Big(\binom{n}{r} - \binom{n-M}{r}\Big)\mu(B)$$

$$= \Big(1 - \Big(\binom{n-m}{r}\Big/\binom{n}{r}\Big)\Big)\mu(B)$$

$$= (1 - (n-M)!(n-r)!/n!(n-r-m)!))\mu(B)$$

$$= (1-((n-r)(n-r-1)\ldots(n-r-M+1)/n(n-1)\ldots(n-M+1)))\mu(B)$$

$$= \Big(1 - \Big(1-\frac{r}{n}\Big)\Big(1-\frac{r}{n-1}\Big)\ldots\Big(1-\frac{r}{n-M+1}\Big)\Big)\mu(B)$$

$$\geq \Big(1 - \Big(1-\frac{r}{n}\Big)^M\Big)\mu(B).$$

From this the statement follows immediately. □

We are now in a position to give the main result of this section.

THEOREM 8.4. (Kwapien [1973]). Let $T : L_0 \to L_0(\mu)$ be a linear operator. Then

$$Tf(x) = \sum_{n=1}^{\infty} g_n(x) f(\sigma_n(x))$$

for every $f \in L_0$, μ a.e., where

(i) each $\sigma_n : \Omega \to \Delta$ is a non-singular Σ-M measurable map,

(ii) each g_n is in $L_0(\mu)$,

(iii) for almost all x in Ω, $g_n(x) \neq 0$ for only finitely many n.

Conversely, every map defined in the above way is a linear operator from L_0 to $L_0(\mu)$.

Proof. For each E in M, define $\Phi(E) = \sup\{\text{supp } Tf : f \in L_0(E)\}$ (the lattice supremum being taken in Σ). Now define

$$g = \sup \{ \sum_{i=1}^{n} 1_{\Phi(E_i)} : E_1, \ldots, E_n \text{ are pairwise disjoint sets in } M \}.$$

From the discussion before Lemma 8.3, g exists as an extended real-valued measurable function, taking on only integer values or possibly the value $+\infty$.

We claim that

(1) If $E_1, E_2 \in M$, then $\Phi(E_1 \cup E_2) = \Phi(E_1) \cup \Phi(E_2)$

(2) If (E_n) is a sequence in M and $\lambda(E_n) \to 0$, then $\mu(\Phi(E_n)) \to 0$, and

(3) $g(x) < \infty$ a.e. μ.

(1) Clearly $\Phi(A) \subset \Phi(B)$ if $A \subset B$. Thus $\Phi(E_1) \cup \Phi(E_2) \subset \Phi(E_1 \cup E_2)$. Now suppose $f \in L_0$ and $\operatorname{supp} f \subset E_1 \cup E_2$. Then $f = f_1 + f_2$ where $f_1 \in L_0(E_1)$ and $f_2 \in L_0(E_2)$. Hence

$$\begin{aligned}\operatorname{supp} Tf &= \operatorname{supp}(Tf_1+Tf_2) \subset \operatorname{supp}(Tf_1) \cup \operatorname{supp}(Tf_2) \\ &\subset \Phi(E_1) \cup \Phi(E_2).\end{aligned}$$

Hence $\Phi(E_1 \cup E_2) \subset \Phi(E_1) \cup \Phi(E_2)$.

(2) Suppose $E_n \in M$ and $\lambda(E_n) \to 0$ but $\mu(E_n) \not\to 0$. By passing to a subsequence if necessary, we may assume that there exists $\epsilon > 0$ such that $\mu(\Phi(E_n)) > \epsilon$ for all n. Thus for each n, there exist $f_1,\ldots,f_k$ in $L_0(E_n)$ such that $\mu(\operatorname{supp} Tf_1 \cup \ldots \cup \operatorname{supp} Tf_k) > \epsilon$. But then by Lemma 8.2 there is a single function g_n in $L_0(E_n)$ (a linear combination of the f_i's) such that $\mu(\operatorname{supp} Tg_n) > \epsilon$. For each n there is $\alpha_n > 0$ such that $\|T(\alpha_n g_n)\|_0 > \epsilon/2$. However, $\|\alpha_n g_n\|_0 \to 0$, and this contradicts Lemma 8.1.

(3) Suppose $g = \infty$ on a set F with $\mu(F) > \alpha > 0$. Then for each $M \in \mathbf{N}$, there is a set $E \in \Sigma(F)$ such that

(i) $\mu(E) > \alpha$, and

(ii) there are sets $E_1,\dots,E_m$, pairwise disjoint, with

$$\sum_{i=1}^{m} 1_{\Phi(E_i)} \geq M 1_E.$$

Since λ is non-atomic, we can refine the sets E_i by a disjoint collection $(F_j)_{j=1}^{n}$ in M so that

$$\bigcup_{j=1}^{n} F_j = \bigcup_{i=1}^{m} E_i \quad \text{and}$$

$$\lambda(F_i) \leq 2\lambda(F_j) \quad \text{for all } i,j.$$

By (1) $\displaystyle\sum_{j=1}^{n} 1_{\Phi(F_j)} = \sum_{i=1}^{m} 1_{\Phi(E_i)} \geq M 1_E.$

Now by Lemma 8.3, given r, $1 \leq r \leq n$, we may assume that

$$\mu(\Phi(\bigcup_{j=1}^{r} F_j)) = \mu(\bigcup_{j=1}^{r} \Phi(F_j)) \geq (1-(1-(r/n))^M)\alpha.$$

Also, $\lambda(F_1 \cup \dots \cup F_r) \leq 2r/n$. Given M, we may choose $r = r(M)$ so that if we set $\epsilon_M = r/n$, then $\lim_{M\to\infty} \epsilon_M = 0$ and also $\lim_{M\to\infty} (1-\epsilon_M)^M = 0$.

But now if we take $G_n = \bigcup_{j=1}^{r} F_j$, we have $\lambda(G_n) < 2\epsilon_M$ and $\mu(\Phi(G_n)) \to \alpha > 0$. This contradicts (2), and shows that g is finite μ-almost everywhere.

Let $B_n = \{x : g(x) \leq n\}$; then $B_1 \subset B_2 \subset \ldots$ and $\mu(\Omega \sim \bigcup_{n=1}^{\infty} B_n) = 0$.

We shall show that there exist sets A_n in Σ, non-singular Σ-M measurable maps $\sigma_{i,n}$, $1 \leq i \leq k_n$, from A_n into Δ, and functions $g_{i,n}$ in $L_0(A_n)$, $1 \leq i \leq k_n$, with the following properties:

(i) the (A_n) are pairwise disjoint,

(ii) $\mu(B_n \sim \bigcup_{i=1}^{n} A_i) < 1/n$,

(iii) $A_n \subset B_n$ for every n,

(iv) $\operatorname{supp} g_{i,n} \subset A_n$, $1 \leq i \leq k_n$,

(v) $1_{A_n} Tf = \sum_{i=1}^{k_n} g_{i,n} f \circ \sigma_{i,n}$, for all f.

This will prove the theorem since then

$$Tf = \sum_{n=1}^{\infty} \sum_{i=1}^{k_n} g_{i,n} f \circ \sigma_{i,n} \quad \text{for all } f.$$

We are going to choose the sets A_n inductively. Suppose that $A_1, \ldots, A_{n-1}$ have been chosen. Now g is finite on B_n, so there exist sets $(E_i)_{i=1}^{k}$ in Σ, pairwise disjoint, such that if

$$F_n = \{x \in B_n : g(x) > \sum_{i=1}^{k} 1_{\Phi(E_i)}\},$$

then $\mu(F_n) < 1/n$. (Recall that g is integer valued.) Let $A_n = B_n \sim (A_1 \cup \ldots \cup A_{n-1} \cup F_n)$. Then

$$\mu(B_n \sim (A_1 \cup \ldots \cup A_n)) \leq \mu(F_n) < 1/n.$$

Suppose that $A \cup B = E_i$ for some i with $A \cap B = \phi$ and $A, B \in M$. We claim that $\mu(\Phi(A) \cap \Phi(B) \cap A_n) = 0$. (The content of this is that functions with disjoint supports in E_i are carried by T to functions with disjoint supports, when restricted to A_n.) Suppose the claim is false. Then

$$\{E_1, \ldots, E_{i-1}, A, B, E_{i+1}, \ldots, E_k\}$$

is a disjoint collection and

$$\sum_{j \neq i} 1_{\Phi(E_j)} + 1_{\Phi(A)} + 1_{\Phi(B)} > g$$

when restricted to $\Phi(A) \cap \Phi(B) \cap A_n$. But this contradicts the definition of F_n. Hence the map $\eta_{i,n} : M(E_i)$ to $\Sigma(A_n)$ defined by

$$\eta_{i,n}(A) = \Phi(A) \cap A_n$$

is a continuous Boolean homomorphism. By our assumption on M

and Σ, there exists $\sigma_{i,n} : A_n \to E_i$, nonsingular, such that $\eta_{i,n}(A) = \sigma_{i,n}^{-1}(A)$ for every A in $M(E_i)$. Let $g_{i,n} = 1_{A_n} T 1_{E_i}$. If A, B are in $M(E_i)$ with $A \cap B = \phi$ and $A \cup B = E_i$, then

$$\operatorname{supp} 1_{A_n} T 1_A \subset A_n \cap \Phi(A)$$

and

$$\operatorname{supp} 1_{A_n} T 1_B \subset A_n \cap \Phi(B).$$

Hence

$$\begin{aligned} 1_{A_n} T(1_A) &= g_{i,n} 1_{\Phi(A)} \\ &= g_{i,n} 1_{\sigma_{i,n}^{-1}(A)} \\ &= g_{i,n} 1_A \circ \sigma_{i,n}. \end{aligned}$$

Thus if f is in $L_0(E_i)$, $1_{A_n} T(f) = g_{i,n} f \circ \sigma_{i,n}$; and finally if $f \in L_0$,

$$1_{A_n} Tf = \sum_{i=1}^{k_n} g_{i,n} f \circ \sigma_{i,n}.$$

This completes the proof. □

<u>REMARK</u>. It is worthwhile singling out the properties of the map Φ used in the proof. Initially, Φ satisfies

(i) $\Phi(A \cup B) = \Phi(A) \cup \Phi(B)$

(ii) $\lambda(A_n) \to 0 \implies \mu(\Phi(A_n)) \to 0$.

The proof of the theorem shows that Ω is a countable disjoint union of sets Y_i satisfying the following: for each i, there are disjoint sets $E_{i,j}$ in Δ such that for each j the map $A \to \Phi(A) \cap Y_i$ is a Boolean σ-homorphism on $E_{i,j}$. In other words, Φ is a "piecewise homomorphism."

The following important result is a straightforward consequence of Theorem 8.4:

THEOREM 8.5. Suppose that $T : L_0 \to L_0(\mu)$ is a non-zero operator. Then there is a set A in $\mathcal{M}$, $\lambda(A) > 0$, such that the restriction of T to $L_0(A)$ is an isomorphism.

<u>Sketch of proof</u>. If T is non-zero, then (cf. the proof of Theorem 8.4) for some n and some $i \leq k_n$,

$$g_{i,n} = 1_{A_n} T 1_{E_i} \neq 0.$$

From the construction, it is easily verified that

$$f \to 1_{A_n} Tf$$

is an isomorphism of $L_0(E_i)$ into $L_0(\operatorname{supp} g_{i,n})$; from this it is immediate that T is an isomorphism when restricted to $L_0(E_i)$. $\square$

REMARK. Theorem 8.4 gives a representation of a linear operator T in the form

$$Tf(x) = \sum_{n=1}^{\infty} g_n(x)\delta_{\sigma_n(x)}(f).$$

It is not hard to see that the expression $\sum_n g_n(\ )\delta_{\sigma_n(\ )}$ is essentially uniquely determined by T.

A further fact about L_0 is:

THEOREM 8.6. (Kalton [1978a]) Every complemented subspace of L_0 is isomorphic to L_0.

The idea of the proof is to show that a complemented subspace of L_0 contains a further complemented subspace which is isomorphic to L_0 and then to use the Pelczynski decomposition method. The first step uses heavily the special form of a projection on L_0.

Remarks. In view of Theorem 8.5 it is natural to ask whether a non-zero operator from L_0 into an arbitrary F-space must be an isomorphism when restricted to $L_0(A)$ for some A of positive measure. In the above paper, Kalton answered this negatively by constructing a quotient space of L_0 such that the quotient map on L_0 fails to have the above property.

*Problem 8.1. Must a non-zero operator from L_0 into an arbitrary F-space preserve a copy of L_0?

3. Operators between L_p spaces, $0 < p < 1$

In this section we give the representation theorem for operators from L_p to $L_p(\mu)$, $0 < p < 1$. As a corollary we show that a non-zero operator from L_p to $L_p(\mu)$ preserves a copy of L_p.

The form of the general operator is the same as in the L_0 case

$$Tf(x) = \sum_{n=1}^{\infty} g_n(x)f(\sigma_n(x)), \tag{8.1}$$

with (g_n), (σ_n) satisfying the appropriate conditions; but the method of proof is quite different. The strategy of the proof is this: we first show that

$$Tf(x) = \int f d\nu_x$$

for a measure ν_x; this is the longest part of the proof and uses a weak-* compactness argument. We then show that for almost all x, ν_x is purely atomic, and from this we derive (8.1). (Contrast the method here with the "support map" technique of the previous section.)

As before, we assume that Σ is μ-complete.

Our first definition and lemma give one way to prove that a measure is purely atomic.

DEFINITION. Let (Ω,Σ,μ) be a measure space. For $0 < p < 1$ and for $A \in \Sigma$, define

$$|\mu|_p(A) = \sup\{\Sigma|\mu(A_i)|^p : A_i \in \Sigma,\ (A_i) \text{ a countable partition of } A\}.$$

It is easy to check that $|\mu|_p$ is a (possibly infinite-valued) non-negative measure on Σ.

We leave the proof of the next lemma to the reader.

LEMMA 8.7. (i) If $|\mu|_p(\Omega) < \infty$ for some $p < 1$, then μ is purely atomic;

(ii) if $\mu = \Sigma\alpha_i\delta_{x_i}$, then

$$|\mu|_p(A) = \sum_{x_i \in A} |\alpha_i|^p, \quad A \in \Sigma.$$

In the proof of the representation theorem, then, we will show that $|\nu_x|_p(\Omega) < \infty$ for almost all x. We will also have to show that $x \to \nu_x$ is, in an appropriate sense, a measurable map.

We are going to prove the representation theorem for an operator T under the assumption that the domain of T is L_p of the Cantor set. This is not a significant loss of generality; the general case is proved in essentially the same way. The assumption does simplify the proofs of some of the measurability lemmas to follow.

The domain of our operators T will be $L_p(\Delta, \mathcal{B}, \lambda)$, where Δ is the Cantor set, $\mathcal{B}$ is the Borel sets of Δ, and λ is normalized Haar measure on Δ. Δ will be thought of as the countable product of the two-points sets $\{0,1\}$. T will map into $L_p(\Omega, \Sigma, \mu)$.

NOTATION. $M(\Delta)$ denotes the regular Borel measures on Δ $(= C(\Delta)^*)$; W^* denotes the σ-algebra generated by the weak-star open sets in $M(\Delta)$. For convenience, we recall the notion of monotone class and the Monotone Class Lemma (See Halmos [1950]).

DEFINITION. A family $\mathcal{C}$ of subsets of a fixed set is a <u>monotone class</u> if

(i) $C_i \in \mathcal{C}$, $C_i \subset C_{i+1}$,

$$i = 1,2,\ldots \text{ implies } \bigcup_i C_i \in \mathcal{C};$$

(ii) $C_i \in \mathcal{C}$, $C_i \supset C_{i+1}$,

$$i = 1,2,\ldots \text{ implies } \bigcap_i C_i \in \mathcal{C}.$$

LEMMA 8.8. (Monotone Class Lemma). Let $\mathcal{C}$ be a monotone class which contains a ring of sets $\mathcal{R}$. Then $\mathcal{C}$ contains the σ-ring generated by $\mathcal{R}$.

LEMMA 8.9. The map $\eta \to |\eta|$ is W*-W* Borel measurable ($|\eta|$ is the total variation of the measure η.)

We leave the proof as an exercise. □

LEMMA 8.10. Suppose $x \to \nu_x$ is a map from Ω to $M(\Delta)$, with $\|\nu_x\| \leq M < \infty$ for all x. Then the following are equivalent:

(i) $x \to \nu_x$ is Σ-W* measurable;

(ii) for all clopen B, $x \to \nu_x(B)$ is Σ-measurable;

(iii) for all Borel B, $x \to \nu_x(B)$ is Σ-measurable.

<u>Proof</u>. (i) <==> (ii)

Sets of the form $\{\eta : \eta(B) \leq K\}$, B clopen, are a subbase for the restriction of W* to any norm-bounded set of measures. This is a simple consequence of the fact that finite linear combinations of characteristic functions of clopen sets are norm-dense in $C(\Delta)$. So

$$x \to \nu_x \text{ is } \Sigma\text{-}W^* \text{ measurable}$$

$$\iff \text{for } B \text{ clopen } \{x : \nu_x(B) \leq K\} \text{ is in } \Sigma.$$

$$\iff x \to \nu_x(B) \text{ is } \Sigma\text{-measurable, for } B \text{ clopen.}$$

(ii) ==> (iii) Let $\mathcal{C} = \{B \in \mathcal{B}: x \to \nu_x(B)$ is Σ-measurable$\}$. Then $\mathcal{C}$ contains the ring of clopen sets. $\mathcal{C}$ is also a monotone class: if $B_n \in \mathcal{B}$, $B_n \subset B_{n+1}$ for all n, then $\nu_x(B_n) \to \nu_x(\bigcup_n B_n)$ for all x. If $B_n \supset B_{n+1}$ for all n, then $\nu_x(B) \to \nu_x(\bigcup_n B_n)$, for all x. By Lemma 8.8, $\mathcal{B} \subset \mathcal{C}$.

LEMMA 8.11. (i) the map $\eta \to |\eta|_p(\Delta)$ is w*-lower semi-continuous; (ii) for U open in Δ, $\eta \to |\eta|_p(U)$ is w*-lower semi-continuous.

__Proof__. $|\eta|_p(\Delta) = \sup\{\sum_{i=1}^n |\eta(C_i)|^p : (C_i)$ is a finite Borel partition of $\Delta\}$. Suppose $\epsilon > 0$. An easy argument using the regularity of η and the fact that clopen sets are a base for the open sets shows that for each i there is a clopen set V_i such that $|\eta|(C_i \Delta V_i) < \epsilon$. Since the C_i's are pairwise disjoint, the V_i's may be taken to be pairwise disjoint. Thus

$$|\eta|_p(\Delta) = \sup\{\sum_{i=1}^n |\eta(V_i)|^p, \ (V_i) \text{ is a clopen partition of } \Delta\}.$$

But the last expression is a supremum of w*-continuous functions of η, so $\eta \to |\eta|_p(\Delta)$ is w*-lower semi-continuous.

The proof of (ii) is the same, with V_i replaced by $V_i \cap U$. □

LEMMA 8.12. Let $x \to \nu_x$ be Σ-W^* measurable and suppose $S_0 \in \Sigma$ is such that $|\nu_x|_p(\Delta) \leq M < \infty$ for x in S_0. Then the map $x \to |\nu_x|_p(B)$ is Σ-measurable, for each B in $\mathcal{B}$.

<u>Proof</u>. Note that $\|\nu_x\| \leq (|\nu_x|_p(\Delta))^{1/p} \leq M^{1/p}$ for x in S_0. By Lemmas 8.10 and 8.11, $x \to |\nu_x|_p(B)$ is Σ-measurable for B clopen; now from Lemma 8.11 again $x \to |\nu_x|_p(B)$ is Σ-measurable for B in $\mathcal{B}$. □

LEMMA 8.13. Suppose S_0 is a set in Σ, and that for each x in S_0 there is a sequence of measures (ν_x^n) so that $x \to \nu_x^n$ is Σ-W^*-measurable, for each n. Suppose $|\nu_x^n|_p(\Delta) \leq M < \infty$, $x \in S_0$, $n = 1,2,\dots$. Finally, suppose $\nu_x^n \longrightarrow \nu_x$ (weak*) for each x in S_0. Then

(i) $x \to \nu_x$ is Σ-W^*-measurable;

(ii) $x \to |\nu_x|_p(B)$ is Σ-measurable for all B in $\mathcal{B}$.

<u>Proof</u>. First, note that $\|\nu_x^n\| \leq M^{1/p}$ for all x in S_0 and all n. Now by Lemma 8.10, $x \to \nu_x$ is Σ-W^* measurable. Also, $|\nu_x|_p(\Delta) \leq M$ for all x in S_0. Lemma 8.12 now implies that $x \to |\nu_x|_p(B)$ is Σ-measurable for each B in $\mathcal{B}$. □

Armed with these facts, we now turn to the representation of an operator from L_p to $L_p(\mu)$.

PROPOSITION 8.14. Suppose that $x \to \nu_x$ is a map from Ω to $M(\Delta)$ satisfying

(i) $x \to \nu_x(B)$ is Σ-measurable for each Borel set B;

(ii) $x \to |\nu_x|_p(B)$ is Σ-measurable for each Borel set B;

(iii) $\sup_{\substack{\lambda(B)>0 \\ B\in\mathcal{B}}} \{1/(\lambda(B)) \int |\nu_x|_p(B)d\mu(x)\} = M < \infty$.

Then $Tf(x) = \int f d\nu_x$ defines a bounded linear operator on the simple functions in $L_p(\lambda)$ into $L_p(\Omega,\Sigma,\mu)$ and $\|T\|^p \leq M$.

<u>Proof</u>. If $f = \sum_{i=1}^{n} \alpha_i 1_{B_i}$, then $Tf(x) = \sum_{i=1}^{n} \alpha_i \nu_x(B_i)$. By (i) Tf is a measurable function of x. Now suppose that f has the above representation, with $B_i \cap B_j = \phi$ if $i \neq j$. Then

$$\|Tf\|^p \leq \sum_{i=1}^{n} |\alpha_i|^p \int |\nu_x(B_i)|^p d\mu(x)$$

$$\leq \sum_{i=1}^{n} |\alpha_i|^p M\lambda(B_i),$$

$$= M\|f\|^p.$$

This proves the claim. Now T extends to a bounded linear operator from $L_p(\lambda)$ to $L_p(\Omega,\Sigma;\mu)$ with norm still at most $M^{1/p}$. □

It may be further established for <u>any</u> $f \in L_p$ that f is $|\nu_x|$-integrable μ almost everywhere and

$$Tf(x) = \int f d\nu_x \qquad \mu\text{-a.e.}$$

See the reference for details.

We now come to the first main result:

THEOREM 8.15. (Kalton [1978a]). Any linear operator $T : L_p(\lambda) \to L_p(\Omega,\Sigma,\mu)$ has the form $Tf(x) = \int f d\nu_x$, f Borel simple, where

(i) $x \to \nu_x(B)$ is Σ-measurable for $B \in \mathcal{B}$;

(ii) $x \to |\nu_x|_p(B)$ is Σ-measurable for $B \in \mathcal{B}$;

(iii) $$\|T\|^p = \sup_{\substack{\lambda(B)>0 \\ B\in\mathcal{B}}} \{(1/(\lambda(B)) \int |\nu_x|_p(B) d\mu(x)\}.$$

<u>Proof</u>. For $n \in \mathbf{N}$, and $1 \leq k \leq 2^n$, write $k-1 = \sum_{i=1}^{n} \epsilon_i 2^{i-1}$, $\epsilon_i = 0$ or 1, and let

$$\Delta_k^n = \{x \in \Delta : x_i = \epsilon_i,\ 1 \leq i \leq n\}.$$

This is a clopen set. Let $w_k^n = T1_{\Delta_k^n}$.

For any $\eta \in M(\Delta)$ and any Σ-measurable $h : X \to \mathbf{R}$, the map $x \to h(x)\eta$ is Σ-W* measurable. It follows that for any choice of $\tau_k^n \in \Delta_k^n$ the map

$$x \to \nu_x^n = \sum_{k=1}^{2^n} w_k^n(x)\delta_{\tau_k^n}$$

is Σ-W*-measurable.

For any n, and $1 \leq k \leq 2^n$, $\Delta_k^n = \Delta_{2k-1}^{n+1} \cup \Delta_{2k}^{n+1}$, and the union is disjoint. Hence there is a μ-null set N_1 such that for $x \notin N_1$, $w_k^n = w_{2k-1}^{n+1} + w_{2k}^{n+1}$, and further

$$(*) \qquad w_k^n(x) = \int 1_{\Delta_k^n} d\nu_x^n$$

$$= \int 1_{\Delta_k^n} d\nu_x^m \quad \text{for all } m \geq n.$$

But

$$|\nu_x^n|_p(\Delta) = \sum_{k=1}^{2^n} |w_k^n(x)|^p,$$

so

$$\int |\nu_x^n|_p(\Delta)d\mu = \sum_{k=1}^{2^n} \int |w_k^n(x)|^p d\mu(x)$$

$$\leq \sum_{k=1}^{2^n} \|T\|^p \|1_{\Delta_k^n}\|^p$$

$$= \|T\|^p.$$

Also,

$$|\nu_x^{n+1}|_p(\Delta) = \sum_{k=1}^{2^{n+1}} |w_k^{n+1}(x)|^p$$

$$\geq |\nu_x^n|_p(\Delta),$$

so by the Monotone Convergence Theorem, $\lim_n |\nu_x^n|^p(\Delta)$ is an integrable function of x. Call this integrable function g, and let N_2 be a μ-null set such that g is finite off N_2. We can now obtain the desired measures ν_x as follows: since

$$\|\nu_x^n\| \leq (|\nu_x^n|_p(\Delta))^{1/p} \leq (g(x))^{1/p}$$

we have that for $x \notin N_2$ the measures ν_x^n are norm-bounded and hence have a w*-cluster point. Call this cluster point ν_x for $x \notin N_1 \cup N_2$, and for $x \in N_1 \cup N_2$ let $\nu_x = 0$.

The measure ν_x is unique, since $\nu_x(\Delta_k^n) = \nu_x^n(\Delta_k^n)$ for all n and k, $1 \leq k \leq 2^n$, by *); therefore (weak*) $\nu_x^n \longrightarrow \nu_x$. To prove measurability, let (Ω_j) be a partition of Ω by sets in Σ such that $g \leq j$ on Ω_j. From Lemma 8.13 applied to each Ω_j, $x \to \nu_x$ is Σ-W* measurable and $x \to |\nu_x|_p(B)$ is Σ-measurable for each $B \in \mathcal{B}$.

Calculating the norm of T in terms of ν_x, we have that for $n \geq m$ and $1 \leq k \leq 2^n$

$$\int_\Delta |\nu_x^n|_p(\Delta_k^m)d\mu \leq \|T\|^p \sum \{\lambda(\Delta_i^n) : \Delta_i^n \subset \Delta_k^m\}$$

$$= \|T\|^p \lambda(\Delta_k^m).$$

Hence

$$\int_\Delta |\nu_x|_p(\Delta_k^m)d\mu \leq \|T\|^p \lambda(\Delta_k^m)$$

by Lemma 8.11, and it then follows that

$$\int_\Delta |\nu_x|_p(B)d\mu \leq \|T\|^p \lambda(B)$$

for all $B \in \mathcal{B}$.

Thus one inequality in (iii) of the statement of the theorem has been proved. To get the reverse inequality, observe that we can define an operator S on the simple functions in $L_p(\Delta,\mathcal{B},\lambda)$ by

$$(Sf)(x) = \int f d\nu_x .$$

From Proposition 8.14,

$$\|S\|^p \le \sup_{\substack{\lambda(B)>0 \\ B\in\mathcal{B}}} \{(1/(\lambda(B)) \int |\nu_x|_p(B)d\mu\}.$$

By construction, S coincides with T, and so the reverse inequality holds, and the proof is complete. □

At this point we have a representation for T as

$$Tf(x) = \int f d\nu_x .$$

We want to derive a representation

$$Tf(x) = \sum_{n=1}^{\infty} g_n(x) f(\sigma_n(x)).$$

The main tool here is a "measurable" version of the fact that a measure can be decomposed into atomic and continuous parts. We need some notation: $M_c(\Delta)$ will denote the set of <u>continuous</u> measures in $M(\Delta)$. $\mathcal{U}^*$ will denote the σ-algebra of universally measurable subsets of $M(\Delta)$. ($\mathcal{U}^*$ is the intersection of all completions of W^* with respect to finite regular Borel measures defined on W^*.)

Now we can state the result we need; we refer the reader to Kalton [1978a] for the proof.

THEOREM 8.16. There exist maps

$$b_n : M(\Delta) \to \mathbf{R}, \qquad n = 1,2,\ldots,$$

$$h_n : M(\Delta) \to \Delta, \qquad n = 1,2,\ldots,$$

$$\varphi \quad : M(\Delta) \to M_c(\Delta),$$

such that

(i) b_n is $\mathcal{U}^*$-measurable,

(ii) h_n is $\mathcal{U}^*$-$\mathcal{B}$ measurable,

(iii) φ is $\mathcal{U}^*$-W^* measurable,

(iv) $|b_n(\eta)| \geq |b_{n+1}(\eta)| \quad n \in \mathbf{N}, \quad \eta \in M(\Delta)$;

(v) $h_n(\eta) \neq h_m(\eta), \quad m \neq n, \quad \eta \in M(\Delta)$,

(vi) $\eta = \sum_n b_n(\eta)\delta_{h_n(\eta)} + \varphi(\eta), \qquad \eta \in M(\Delta)$.

THEOREM 8.17. Let $(\Delta,\mathcal{B},\lambda)$ and (Ω,Σ,μ) be as before. Then for $p < 1$, every linear operator $T : L_p(\Delta,\mathcal{B},\lambda) \to L_p(\Omega,\Sigma,\mu)$ has the form

$$Tf(x) = \sum_n g_n(x)f(\sigma_n(x)) \qquad \mu \text{ a.e.},$$

where

(i) g_n is Σ-measurable,

(ii) σ_n is Σ-$\mathcal{B}$ measurable,

(iii) $|g_n(x)| \geq |g_{n+1}(x)|, \quad x \in \Omega, \quad n \in \mathbf{N}$,

(iv) $\sigma_n(x) \neq \sigma_m(x), \quad m \neq n, \quad x \in \Omega$,

(v) $\sum_n |g_n(x)|^p < \infty \quad \mu$ a.e., $x \in \Omega$,

(vi) $\sum_n \int_{\sigma_n^{-1}(B)} |g_n(x)|^p d\mu(x) \leq \|T\|^p \lambda(B), \quad B \in \mathcal{B}$.

REMARK. Our proof is only valid for f simple, but the theorem holds as stated.

<u>Proof</u>. Define $g_n(x) = b_n(\nu_x)$ and $\sigma_n(x) = h_n(\nu_x)$, where $x \to \nu_x$ represents T. (b_n and σ_n are as in Theorem 8.16.)

The only problem is the possible lack of measurability of g_n and σ_n, since we do know that

$$\nu_x = \sum_n g_n(x)\delta_{\sigma_n(x)} \quad \mu \text{ a.e. .}$$

Define a measure γ on W^* by

$$\gamma(C) = \mu(\nu^{-1}(C)) = \mu\{x : \nu_x \in C\}.$$

Let $B \in \mathcal{B}$ and $n \in \mathbf{N}$; then $h_n^{-1}(B)$ is γ-measurable, so there exist C_1, C_2 in W^* such that $C_1 \subset h_n^{-1}(B) \subset C_2$ and $\gamma(C_2 \sim C_1) = 0$. Then $\nu^{-1}(C_1) \subset \sigma_n^{-1}(B) \subset \nu^{-1}(C_2)$ and $\mu(\nu^{-1}(C_2) \sim \nu^{-1}(C_1)) = 0$. This shows that σ_n is Σ-$\mathcal{B}$ measurable since Σ is complete.

Similarly, g_n is Σ-measurable. Thus (i) and (ii) of the theorem are satisfied. Statements (iii)-(vi) of the theorem follow easily from the properties of b_n and h_n in Theorem 8.16 and from statements (i)-(iii) of Theorem 8.15. □

We illustrate the use of the representation theorem by showing that a non-zero operator on $L_p(\lambda)$ into $L_p(\Omega,\Sigma,\mu)$ is an isomorphism on $L_p(B)$ for some B in $\mathcal{B}$ of positive measure. The proof uses a simple approximation argument. We first gather the few necessary facts.

For the setting of the next lemma, suppose σ is a Σ-measurable map from Ω to Δ and g is a real-valued Σ-measurable function on Ω.

LEMMA 8.18. Assume g satisfies

$$(*) \qquad \int_{\sigma^{-1}(B)} |g|^p d\mu \leq K\lambda(B)$$

for some $K > 0$ and all B in $\mathcal{B}$. Define $S : L_p(\lambda) \to L_p(\mu)$ by $(Sf)(x) = g(x)f(\sigma(x))$, and assume S is a non-zero operator. Then there is a set B_0 in $\mathcal{B}$ of positive measure such that S is an isomorphism on $L_p(B_0)$.

Proof. Define a measure η on $\mathcal{B}$ by $\eta(B) = \int_{\sigma^{-1}(B)} |g|^p d\mu$. Then since S is non-zero $\eta(\Delta) > 0$; further $\eta << \lambda$ by $(*)$. Let h be the Radon-Nikodym derivative of η with respect to λ. Then there are $\epsilon > 0$ and a set B_0 of positive measure such that $h \geq \epsilon$ on B_0. We claim S is an isomorphism on $L_p(B_0)$. Indeed, if $f \in L_p(B_0)$,

$$\begin{aligned}\int |g(x)f(\sigma(x))|^p d\mu &= \int |f|^p d\eta \\ &= \int |f|^p h d\lambda \\ &\geq \epsilon \|f\|^p. \qquad \square\end{aligned}$$

LEMMA 8.19. Let $T : L_p(\lambda) \to L_p(\mu)$ be a linear operator. Let $x \to \nu_x$ represent T as in 8.15, and suppose

$$\int |\nu_x|_p(\Delta)d\mu < \epsilon.$$

Let B_0 be any set in $\mathcal{B}$ of positive measure. Then there is a subset C of B_0, $C \in \mathcal{B}$, $\lambda(C) > 0$, such that the restriction of T to $L_p(C)$ has norm at most $\epsilon^{1/p}$.

<u>Proof</u>. Define a measure η on $\mathcal{B}$ by

$$\eta(B) = \int |\nu_x|_p(B \cap B_0)d\mu(x).$$

From (iii) of Theorem 8.15, $\eta << \lambda$. Let h be the Radon-Nikodym derivative of η with respect to λ. Since $\eta(B_0) < \epsilon$, it follows that there is a subset C of B_0, $\lambda(C) > 0$, such that $h < \epsilon$ on C.

Now if $B \in \mathcal{B}$, $B \subset C$, then

$$\int |\nu_x|_p(B)d\mu(x) = \eta(B)$$

$$< \epsilon \cdot \lambda(B).$$

Let $\nu_x'(B) = \nu_x(B \cap C)$; then $x \to \nu_x'$ represents the restriction of T to $L_p(C)$. By Theorem 8.15 again and the above, the norm of this restriction is at most $\epsilon^{1/p}$ and the proof is complete. □

We can now prove the promised result:

THEOREM 8.20. Let $T : L_p(\lambda) \to L_p(\mu)$ be a non-zero operator, $p < 1$. Then there is a set C_0 in $\mathcal{B}$ of positive measure such that the restriction of T to $L_p(C_0)$ is an isomorphism.

Proof. Use Theorem 8.16 to write T in the form

$$Tf(x) = \sum_n g_n(x)f(\sigma_n(x)).$$

Then for some n_0, g_{n_0} is not identically zero. Let C be a subset of Ω, $\mu(C) > 0$, such that $|g_{n_0}(x)| > 0$, $x \in C$. For a set A define the operator $f \to 1_A \cdot f$ by R_A.

For each $m \in \mathbf{N}$ let (Δ_i^m) be the mth dyadic partition of Δ as in the proof of Theorem 8.15, and for each i let

$$C_i^m = \sigma_{n_0}^{-1}(\Delta_i^m) \cap C.$$

If $x \to \nu_x$ represents T, then $x \to \nu_x^m$ represents the operator U_m, where

$$U_m = \sum_{k=1}^{2^m} R_{C_i^m} T R_{\Delta_i^m}$$

and

$$\nu_x^m(B) = \begin{cases} \nu_x(B \cap \Delta_i^m), & x \in C_i^m \\ 0, & \text{otherwise.} \end{cases}$$

Let S be defined by

$$Sf(x) = \begin{cases} g_{n_0}(x)f(\sigma_{n_0}(x)), & x \in C \\ 0, & x \notin C. \end{cases}$$

Then $x \to \mu_x$ represents S, where

$$\mu_x = \begin{cases} g_{n_0}(x)\delta_{\sigma_{n_0}}(x), & x \in C \\ 0, & x \notin C. \end{cases}$$

Now if $x \in C_i^m$, then

$$|\mu_x - \nu_x^m|_p(\Delta) = |\mu_x - \nu_x|_p(\Delta_i^m).$$

If $x \in C$, let $i(m)$ be such that $x \in C_{i(m)}^m$. Then

$$\begin{aligned} \lim_{m\to\infty} |\mu_x - \nu_x^m|_p(\Delta) &= \lim_{m\to\infty} |\mu_x - \nu_x|_p(\Delta_{i(m)}^m) \\ &= |\mu_x - \nu_x|_p(\sigma_{n_0}(x)) \\ &= 0; \end{aligned}$$

here we have used (iv) and (v) of Theorem 8.17.

By Lemma 8.18 there is a set B_0 of positive measure such that the restriction of S to $L_p(B_0)$ is an isomorphism on $L_p(B_0)$. Now by Proposition 7.8 and Lemma 8.19 there are $m_0 \in \mathbf{N}$ and a subset B_1 of B_0 with $\lambda(B_1) > 0$ such that the restriction of U_{m_0} to $L_p(B_1)$ is an isomorphism.

Finally, for some i, $1 \leq i \leq 2^{m_0}$, $\lambda(B_1 \cap \Delta_i^{m_0}) > 0$, and if we set $C_0 = B_1 \cap \Delta_i^{m_0}$, it follows that the restriction of T to $L_p(C_0)$ is an isomorphism. The proof is complete. □

The final theorem of the chapter is more specialized in character. As motivation, note that for $0 < p < 1$ the inclusion map of L_p in L_0 is not an isomorphism on $L_p(A)$ for any set A of positive measure; indeed, it is not an isomorphism on any subspace of L_p which is isomorphic to L_p (see Kalton [1980a].) We _can_ ask whether a non-zero operator from L_p into L_0 preserves a copy of L_q for $p < q < 1$.

Theorem 8.22 answers this affirmatively; it first appeared in the above paper of Kalton with a different proof.

We need the notion of weak L_p and Nikishin's factorization theorem.

DEFINITION. $L_{p,\infty}(\mu) =$

$$\{f \in L_0(\mu) : \sup_{c>0} (c(\mu\{|f| > c\})^{1/p} < \infty\}.$$

The supremum in the above definition is a quasi-norm on $L_{p,\infty}(\mu)$. Kalton [1980a] shows that if $p < 1$, then $L_{p,\infty}(\mu)$ is p-convex.

THEOREM 8.21. (Nikishin [1972]). Let X be a quasi-Banach space of type p, $0 < p \leq 2$, and let $T : X \to L_0(\mu)$ be a linear operator. Then for every $\epsilon > 0$ there is a set R_ϵ in Σ such that $\mu(R_\epsilon) < \epsilon$ and such that $1_{\Omega \sim R_\epsilon} T$ is a linear operator from X into $L_{p,\infty}(\mu)$.

Expositions of this important theorem can be found in Kalton [1980a] and Maurey [1974].

THEOREM 8.22. Let $0 < p < 1$ and let $T : L_p \to L_0$ be a non-zero linear operator. Then for $1 > q > p$ there is a subspace F of L_p which is isomorphic to L_q such that T restricted to F is an isomorphism.

Proof. We give a condensed proof. By Nikishin's theorem, there is an everywhere positive function g_1 such that the operator g_1T $(g_1Tf(x) = g_1(x)(Tf(x))$ is an operator from L_p into $L_{p,\infty}$. Again by Theorem 8.21, there is an everywhere positive g_2 such that g_2T is an operator from L_1 into $L_{1,\infty}$. Let $g = g_1 \wedge g_2$.

Now choose r and s, $p < r < s < q < 1$. Then by the Marcinkiewicz interpolation theorem (Bergh, Lofstrom [1976]), gT is an operator from L_r to L_r and L_s to L_s, and gT is a non-zero operator on L_r and on L_s. By Theorem 8.20 applied twice, there is a set A of positive measure such that the restriction of gT to $L_r(A)$ is an isomorphism for the L_r topology and the restriction of gT to $L_s(A)$ is an isomorphism for the L_s topology.

Next, there is a subspace F of $L_s(A)$ such that F is isomorphic to L_q and the L_s-topology and the topology of convergence in measure coincide on F (i.e., F is "strongly embedded" (Garling [1977]). From the properties of gT it follows that the L_r and L_s topologies coincide on $gT(F)$;

hence $gT(F)$ is strongly embedded. Finally, since multiplication by $1/g$ is a linear homeomorphism of L_0, it follows that T is an isomorphism of F (with the L_p topology) into L_0. □

REMARKS. 1. For $0 \leq p < 1$ those sub-σ-algebras $\mathcal{A}$ of $\mathcal{B}$ such that $L_p(\Delta, \mathcal{A}, \lambda)$ is complemented in $L_p(\Delta, \mathcal{B}, \lambda)$ have been completely classified - see Kalton [1978a].

2. There is a quotient space of $L_p(\lambda)$, $0 < p < 1$, which fails to contain a copy of ℓ_p (let alone a copy of $L_p(\lambda)$) - see Kalton [1980a].

3. It is known (Kalton [1978a]) that for $0 < p < 1$, $L_p(\lambda)$ is <u>primary</u>. That is, if $L_p(\lambda) = E \oplus F$, then at least one of the spaces E, F is isomorphic to L_p.

4. <u>Open problem</u>. Is every complemented subspace of $L_p(\lambda)$, $0 < p < 1$ isomorphic to $L_p(\lambda)$? For L_1, the conjecture is that every complemented subspace of L_1 is isomorphic either to L_1 or ℓ_1. L_1 is known to be primary (Enflo-Starbird [1979]).

CHAPTER 9
COMPACT CONVEX SETS WITH NO EXTREME POINTS

1. Preliminary Remarks

In contrast with the locally convex situation, very little is known about compact convex sets in non-locally convex spaces - as a beginning point we state three problems posed by Klee ([1960], [1961b]), that were open in 1961.

PROBLEM 9.1. If K is a compact convex set and $f : K \to K$ is continuous, does f have a fixed point, i.e. does K have the fixed point property?

PROBLEM 9.2. If K is a compact convex set, is K the closed convex hull of its extreme points and, in particular, does K have any extreme points?

PROBLEM 9.3. If K is a compact convex set we shall say that K is locally convex if every point in K has a base of convex neighborhoods (not necessarily open) in K. Is every compact convex set locally convex?

Problem 9.1 was actually posed much earlier by Schauder in the Scottish Book (problem 54) and is still open. Problems 9.2 and 9.3 have been solved negatively. However, the solution of these two problems was motivated by some positive initial results. If K_1 and K_2 are two compact convex sets and $T : K_1 \to K_2$, T is said to be an affine map if for every $x,y \in K_1$ and $\alpha \in [0,1]$, $T(\alpha x + (1-\alpha)y) = \alpha T(x) + (1-\alpha)T(y)$. If, in addition, T is a homeomorphism from K_1 onto K_2 we say that K_1 and K_2 are affinely homeomorphic. If K is a compact convex set that is affinely homeomorphic to a compact convex set in a locally convex space, we say that K is _embeddable in a locally convex space_. Jamison, O'Brien, and Taylor [1976] and Roberts [1978] independently showed that if K_1 is situated in a locally convex space and $T : K_1 \to K_2$ is a continuous affine map onto K_2, then K_2 is embeddable in a locally convex space. Using this result, Jamison, O'Brien and Taylor showed that if K is a compact convex set such that every point has a neighborhood base consisting of _open_ convex sets in K, then K is affinely embeddable in a locally convex space. Later, Lawson [1976] and Roberts [1978] independently showed that a compact convex set is locally convex if and only if it is embeddable in a locally convex space. In the light of this, Klee's Problem 9.3 really asks whether there are any compact convex sets structurally different from those appearing in locally convex spaces. In Roberts [1977], an F-space was constructed containing a compact convex set with no extreme points, thus solving problems 9.2 and 9.3. Later in Roberts [1975-76], it was shown that the spaces L_p, $0 \leq p < 1$, contain compact convex sets with no extreme points.

The key idea is the notion of a needle point.

DEFINITION. If K_1 and K_2 are compact sets in an F-space $(X, \|\cdot\|)$ and $x \in X$, $d(x,K_1) = \inf\{\|x-y\| : y \in K_1\}$. We let

$$D(K_1;K_2) = \sup\{d(x,K_1) : x \in K_2\} .$$

Notice that $\bar{D}(K_1,K_2) = \max\{D(K_1;K_2), D(K_2;K_1)\}$ is the usual _Hausdorff metric_ on the compact subsets of X. If $x,y \in X$, we let $[x,y]$ denote the line segment from x to y, i.e., $[x,y] = \{\alpha x + (1-\alpha)y : \alpha \in [0,1]\}$.

DEFINITION. Let $(X,\|\cdot\|)$ be an F-space. If $x \in X$, $\epsilon > 0$, and F is a finite subset of X, we say that F is an ϵ-<u>needle set</u> for x if

(1) $x \in \text{co } F$
(2) $y \in F$ implies $\|y\| < \epsilon$
(3) $D([0,x];\text{co}(F \cup \{0\})) < \epsilon$.

If x possesses an ϵ-needle set for every $\epsilon > 0$, then x is called a <u>needle point</u>. If every point in X is a needle point, we say that X is a <u>needle point space</u>.

REMARKS. (1) If X is a needle point space then X has trivial dual. Indeed, X having trivial dual is equivalent to the following: if $x \in X$ and $\epsilon > 0$ then there exists a finite set satisfying conditions (1) and (2). Thus condition (3) is the essential feature. In addition to assuming the existence of a finite set F of small elements with a convex combination "straying" to the point x, we insist that F can be chosen so that the "straying" of convex combinations is highly discreet: $\text{co}(F \cup \{0\})$ also must "hug" the line segment $[0,x]$.

(2) In condition (3) $\text{co } F$ could be used as well as $\text{co}(F \cup \{0\})$; however, $\text{co}(F \cup \{0\})$ is slightly more useful when applying the definition. Note that $\text{co}(F \cup \{0\})$ consists of all subconvex combinations of members of F, i.e., if $F = \{x_1,\ldots,x_n\}$, $\text{co}(F \cup \{0\})$ is the set of all

$$\sum_{i=1}^{n} \alpha_i x_i \quad \text{where} \quad \alpha_i \geq 0 \quad \text{and} \quad \sum_{i=1}^{n} \alpha_i \leq 1.$$

(3) A point x is still a needle point if for every $\epsilon > 0$ we can find a finite set F satisfying conditions (2) and (3) as well as

(1') there exists $y \in \text{co } F$ such that $\|x-y\| < \epsilon$.

To see this, notice that if $F = \{x_1,\ldots,x_n\}$, then $F' = \{x_1+(x-y),\ldots,x_n+(x-y)\}$ is a 2ϵ-needle set for x.

(4) The set of needle points is closed. If (x_n) is a sequence of needle points and $x_n \to x$, it is easy to see that for every $\epsilon > 0$, there exist finite sets F satisfying (1'), (2) and (3).

(5) If X and Y are F-spaces, $T \in L(X,Y)$ and x is a needle point in X, then $T(x)$ is a needle point in Y. This follows directly from the definition.

In section 2 we shall give an elementary example of a compact convex set that is not locally convex and then we shall prove that every needle point space contains a compact convex set with no extreme points. In section 3 we prove that the spaces L_p, $0 < p < 1$, are needle point spaces. Motivated by these results, J. H. Shapiro [1977] asked the following questions: if S_μ is a singular inner function so that $S_\mu H^p$ is weakly dense in H^p $(0 < p < 1)$, does the space $H^p/S_\mu H^p$ contain compact convex sets with no extreme points? It turns out that for at least some singular inner function S_μ, the space $H^p/S_\mu H^p$ is a needle point space (Roberts [to appear]). We shall also prove this in section 3. In section 4 we shall discuss some of the major questions that remain open.

At this point we can show that needle points occur in a rather natural way. Let $(X, \|\cdot\|)$ denote the Ribe space (cf. Chapter 5). For the notation used here see section 4 of Chapter 5. Let $x_0 = (-1,0)$ and for $n \geq 2$, define $F = \{(0,e_1)/\log n, \ldots, (0,e_n)/\log n\}$. It is easily verified that F satisfies conditions (1'), (2) and (3) for $\epsilon > 0$ if $(1/\log n) < \epsilon$. In fact, the points in Rx_0 should be regarded as needle points of a very special nature. Denote the standard norm on ℓ_1 by $\|\cdot\|_1$. Since $X/Rx_0 \cong \ell_1$, we regard $\|\cdot\|_1$ as a pseudonorm on X. If B_ϵ is the open ball of radius $\epsilon > 0$ in X and $x \in B_\epsilon$, then $\|qx\|_1 < \epsilon$, where q is the quotient map. Hence if $x \in \text{co } B_\epsilon$, $\|qx\|_1 < \epsilon$, i.e., there exists a constant $c \in R$ such that $\|x-cx_0\| < \epsilon$. Thus $\text{co } B_\epsilon$ is a sort of "cylinder" about the line Rx_0 of radius ϵ, and the convex hull of <u>any</u> finite set of small elements in X tends to "hug" the line Rx_0.

One way to show that the L_p spaces $(0 \leq p < 1)$ are needlepoint spaces is to show that the Ribe space isomorphically embeds in L_p and then apply the fact that L_p is transitive. In Kalton-Peck [1980] it was shown that a rather large class of function spaces are needlepoint spaces. It was also shown there that the construction we will give in this chapter can be carried out inside the unit ball of L_1. Bourgain and Rosenthal [1980] carried these ideas further and constructed a subspace of L_1 which fails to have the Radon-Nikodym property and whose unit ball is compact for the L_p topology, $0 \leq p < 1$. Using the ideas from Bourgain-Rosenthal, Kalton [1981b] showed that the Ribe space does in fact embed in L_p, $0 \leq p < 1$.

2. Needle Point Spaces Contain Compact Convex Sets with No Extreme Points.

First we shall require a very simple result.

LEMMA 9.1. Let $(X,\|\cdot\|)$ be an F-space.

(1) If A,B and C are compact subsets of X, then

$$D(A;C) \leq D(A;B) + D(B;C).$$

(2) If (E_n) is an increasing sequence of compact sets and $\sum_{n=1}^{\infty} \epsilon_n < \infty$ with $D(E_n;E_{n+1}) < \epsilon_{n+1}$ for each n then $\overline{\cup E_n}$ is compact. Also if $x \in E_1$ implies $\|x\| < \epsilon_1$ then $y \in \cup E_n$ implies $\|y\| < \sum_{n=1}^{\infty} \epsilon_n$.

(3) If E and F are compact convex sets, then

$$D(E;F) = D(E;co(E \cup F)).$$

Proof. (1) is verified easily. To obtain (2), let $\epsilon > 0$. Choose N so that

$$\sum_{n=N+1}^{\infty} \epsilon_n < \epsilon/2.$$

Since E_N is compact, there exists F a finite subset of E_N such that if $x \in E_N$, there exists $y \in F$ such that $\|x-y\| < \epsilon/2$. By (1) if $x \in E_n$ for $n \geq N+1$, there exists $x' \in E_N$ so that $\|x-x'\| < \epsilon_{N+1} + \dots + \epsilon_n$ i.e., $D(E_N;E_n) < \epsilon_{N+1} + \dots + \epsilon_n$. Hence, if $y \in F$ so that

$\|x'-y\| < \epsilon/2$, then $\|x-y\| < \epsilon$, i.e., $\bigcup_{n=1}^{\infty} E_n$ is totally bounded. The second statement in (2) also follows from (1).

To prove (3) notice that $D(E;F) \leqslant D(E;co(E \cup F))$. Since E and F are compact convex sets, it is easily shown that

$$\overline{co}(E \cup F) = co(E \cup F)$$
$$= \{\alpha x+(1-\alpha)y : x \in E, y \in F, \alpha \in [0,1]\}.$$

Let $x \in E$, $y \in F$ and $\alpha \in [0,1]$. There exists $x' \in E$ such that $d(x',y) = d(E,y) \leqslant D(E;F)$. Thus

$$\|\alpha x+(1-\alpha)x'-(\alpha x+(1-\alpha)y)\| = \|(1-\alpha)(x'-y)\| \leqslant D(E;F).$$

Hence $D(E;co(E \cup F)) \leqslant D(E;F)$. □

THEOREM 9.2. The Ribe space contains a compact convex set that is not locally convex.

<u>Proof</u>. Let $\|\cdot\|$ be an F-norm on the Ribe space X, let $x_0 = (-1,0)$ and suppose (ϵ_n) is a positive sequence such that $\sum_{n=1}^{\infty} \epsilon_n < \infty$. Since x_0 is a needle point, for each integer n we can select an ϵ_n-needle set F_n for x_0. Now let

$K_n = co(F_1 \cup \ldots \cup F_n \cup \{0\})$ and let $K = \bigcup_{n=1}^{\infty} K_n$.

By Lemma 9.1,

$$D(K_n;K_{n+1}) = D(K_n;co(K_n \cup co(F_{n+1} \cup \{0\}))$$

$$= D(K_n;co(F_{n+1} \cup \{0\}))$$

$$\leqslant D([0,x_0];co(F_{n+1} \cup \{0\})) < \epsilon_{n+1}.$$

Hence $\overline{K}$ is a compact convex set and since $x_0 \in \text{co } F_n$ with each $\overline{F}_n \subset \overline{K}$, $\overline{K}$ is not locally convex. □

NOTE. The above argument shows that any space containing a nonzero needle point also contains a compact convex set that is not locally convex.

DEFINITION. If $(X, \|\cdot\|)$ is an F-space, K is a compact convex subset of X and $\epsilon > 0$, we say that K is ϵ-<u>generated</u> if there exists a finite number of compact convex sets $K_1, \ldots, K_n$ in K so that

(1) if $x \in K_i$, $1 \leq i \leq n$, then $\|x_i\| < \epsilon$,

(2) $K = \text{co}\{K_1, \ldots, K_n\}$.

LEMMA 9.3. If $(X, \|\cdot\|)$ is an F-space and K is a compact convex subset of X that is ϵ-generated for every $\epsilon > 0$, then $x \in \text{ex } K$ implies $x = 0$.

<u>Proof</u>. Let $x \in \text{ex } K$ and let $\epsilon > 0$. Let $K_1, \ldots, K_n$ be compact convex sets in the above definition. It is easily shown that

$$K = \text{co}(K_1 \cup \ldots \cup K_n)$$
$$= \{\sum_{i=1}^{n} \alpha_i x_i : x \in K_i,\ \alpha_i \geq 0 \text{ and } \sum_{i=1}^{n} \alpha_i = 1\}.$$

Hence $x = \sum_{i=1}^{n} \alpha_i x_i$. Since $x \in \text{ex } K$, $x = x_i$ for some i and therefore $\|x\| < \epsilon$. Since $\epsilon > 0$ is arbitrary, $x = 0$. □

THEOREM 9.4. If X is a needle point space, then X contains a compact convex set with no extreme points.

__Proof__. Our first observation is that it will suffice to construct a compact convex set $K \neq \{0\}$ so that K is ϵ-generated for every $\epsilon > 0$, since then $K-K$ will have no extreme points. To see this notice that $\mathrm{ex}(K-K) \subset \mathrm{ex}\,K - \mathrm{ex}\,K \subset \{0\} - \{0\} = \{0\}$. Thus 0 is the only possible extreme point of K but since $K-K$ is symmetric 0 is not an extreme point of $K-K$.

We construct K inductively. Let $x_0 \neq 0$ and let $\epsilon_n > 0$ with $\sum \epsilon_n < \infty$. We select a sequence $F_0, F_1, \ldots$ of finite sets as follows: let $F_0 = \{x_0\}$, for each integer n let $p_n = |F_n|$, let $F_{n+1}(x)$ be an (ϵ_{n+1}/p_n)-needle set for each $x \in F_n$ and let $F_{n+1} = \cup \{F_{n+1}(x) : x \in F_n\}$. Finally, let $E = \bigcup_{n=0}^{\infty} F_n$ and let $K = \overline{\mathrm{co}}(E \cup \{0\})$. We claim that K is a compact convex set that is ϵ-generated for every $\epsilon > 0$.

If $x \in F_k$ define $D_n(x)$ inductively as follows: $D_n(x) = \emptyset$ if $n < k$, $D_n(x) = \{x\}$ if $n = k$ and $D_n(x) = \cup \{F_n(y) : y \in D_{n-1}(x)\}$ if $n > k$ ($D_n(x)$ should be thought of as the descendants of x at the n'th generation.) Finally let $D(x) = \cup D_n(x)$. Fix a nonnegative integer k. If $n \geq k$ then it is easily seen that $F_n = \bigcup_{x \in F_k} D_n(x)$. Thus $E = \bigcup_{x \in F_k} D(x)$. For $x \in F_k$, let $C_k(x) = \overline{\mathrm{co}}(D_k(x) \cup \{0\})$. We will show that $C_k(x)$ is a compact convex set and if $y \in C_k(x)$, then $\|y\| < \sum_{n=k-1}^{\infty} \epsilon_n$. This will complete the proof since it is then obvious that $K = \mathrm{co}\{C_k(x) : x \in F_k\}$. For $x \in F_k$, let $K_n(x) = \mathrm{co}(D_n(x) \cup \{0\})$ if $n \geq k$. For any $n \geq k$, let $D_n(x) = \{x_1, \ldots, x_m\}$. Note that $m \leq p_n$. Also let

$K_{n_0} = K(x)$ and for $1 \leq i \leq m$, let

$K_{n_i} = co(K_n(x) \cup F_{n+1}(x_1) \cup \ldots \cup F_{n+1}(x_i))$. Now

$$\begin{aligned} D(K_{n_i}(x);K_{n_{i+1}}(x)) &= D(K_{n_i}(x);co(K_{n_i}(x) \cup co(F_{n+1}(x_i))) \\ &= D(K_{n_i}(x);co(F_{n+1}(x_{i+1}))) \\ &\leq D([0,x_{i+1}];co(F_{n+1}(x_{i+1}))) \\ &< \epsilon_{n+1}/p_n. \end{aligned}$$

Thus $D(K_n(x);K_{n+1}(x)) < m\epsilon_{n+1})/p_n \leq \epsilon_{n+1}$.
Since $C_k(x) = \overline{\bigcup_{n=k} K_n(x)}$, $K_k(x) = \{x\}$ and $\|x\| < \epsilon_{k-1}$, by Lemma 9.1 $C_k(x)$ is compact and if $y \in C_k(x)$, $\|y\| < \sum_{n=k-1}^{\infty} \epsilon_n$. □

3. Needle Point Spaces

The first order of business will be to show that for $0 \leq p < 1$, L_p is a needle point space. Notice that it will suffice to show that for $0 < p < 1$, the constant function 1 is a needle point in L_p, for the inclusion map of L_p into L_0 is continuous and we then have that 1 is a needle point in L_0. By transitivity the spaces L_p, $0 \leq p < 1$, are needle point spaces. For $0 < p < 1$, we let $\|\cdot\|$ denote the standard p-norm on L_p, i.e. if $f \in L_p$,

$$\|f\| = \|f\|_p^{\,p} = \int_0^1 |f(x)|^p dx.$$

Notice that if $1 \le q \le \infty$ (we shall only be interested in $q = 1$ or 2) then $L_q \subset L_p$ and if $f \in L_q$ then $\|f\|_p \le \|f\|_q$ so that

$$(9.1) \qquad \|f\| \le \|f\|_q^{\,p}.$$

We shall now assume some very basic results from probability (for details see any graduate text book). Much of the motivation here comes from the law of large numbers. Recall that if $f \in L_2$ then the <u>variance of</u> f is defined by

$$\text{Var } f = \int_0^1 (f(x) - \int_0^1 f(t)dt)^2 dx = \|f-E(f)\|_2^2$$

where

$$E(f) = \int_0^1 f(t)dt.$$

If $f_1,\ldots,f_n$ is an independent finite sequence in L_2, then

$$\text{Var}(\sum f_i) = \sum \text{Var } f_i.$$

For purposes of motivation we might attempt to show that 1 is a needle point in L_p in the following way: let (f_n) be an independent identically distributed (i.i.d.) sequence in L_2 such that $f_1 \ge 0$, $\int_0^1 f_1(x)dx = 1$. Suppose also that $\alpha_1,\ldots,\alpha_n \ge 0$ and $\sum \alpha_i \le 1$ with $\alpha = \max\{\alpha_1,\ldots,\alpha_n\}$. Then

$$(9.2) \qquad \|\sum_{i=1}^n \alpha_i f_i - \sum_{i=1}^n \alpha_i\|$$

$$\le \|\sum_{i=1}^n (\alpha_i f_i - \alpha_i)\|_2^{\,p}$$

$$= \left(\sum_{i=1}^{n} \alpha_i^2 \operatorname{Var} f_i \right)^{p/2}$$

$$= \left(\operatorname{Var} f_1 \sum_{i=1}^{n} \alpha_i^2 \right)^{p/2}$$

$$\leq (\alpha \operatorname{Var} f_1)^{p/2}.$$

Hence by (9.2) (or by the Law of Large Numbers)

$$1/n \sum_{i=1}^{n} f_i \to 1 \quad \text{in} \quad L_p.$$

Thus if we choose (f_n) with $\|f_1\|$ small and then n suitably large, $F = \{f_1, \ldots, f_n\}$ could be made to satisfy conditions (1') and (2) in the alternate definition of ϵ-needle set (for some pre-given $\epsilon > 0$). The difficulty comes from condition (3). We need to show that if $\alpha_1, \ldots, \alpha_n \geq 0$ and $\Sigma\alpha_i \leq 1$ then $\Sigma\alpha_i f_i$ is approximately (in L_p) a constant in $[0,1]$. Certain subconvex combinations present no problems. If every nonzero α_i is large, there are not many such α_i and $\Sigma\alpha_i f_i$ is small since $\|f_i\| = \|f_1\|$ is small. If each α_i is small, $\Sigma\alpha_i f_i$ is close to $\Sigma\alpha_i$ by (9.2). If $\alpha_i, \ldots, \alpha_n$ consisted of only very large and very small terms we would still have no problems: it is the terms of moderate size that create difficulty. To isolate this problem we introduce the following notation:

DEFINITION. If (f_n) is an i.i.d. sequence in L_2, $f_1 \geq 0$, $\int_0^1 f_1(x)dx = 1$ and $\delta > 0$ then $[a,b] \subset (0,1)$ is called a δ-divergent zone for (f_n) if whenever $\alpha_1,\ldots,\alpha_n \geq 0$ and $\sum_{i=1}^{n} \alpha_i \leq 1$ we have

(1) $\|\Sigma\{\alpha_i f_i : \alpha_i > b\}\| < \delta$

(2) $\|\Sigma\{\alpha_i f_i - \alpha_i : \alpha_i < a\}\| < \delta$.

Note that if (f_n) has a δ-divergent zone, then $\|f_1\| < \delta$. Our approach will be to take the average of several i.i.d. sequences in L_2 with pairwise disjoint δ-divergent zones. This will allow us to "average away" our problems.

LEMMA 9.5. (1) If $\delta,\epsilon \in (0,1)$ then there exists (f_n) an i.i.d. sequence in L_2 such that $f_1 \geq 0$, $\int_0^1 f_1(x)dx = 1$, and (f_n) has a δ-divergent zone $[a,b] \subset (0,\epsilon)$.

(2) If k is a positive integer and $\delta > 0$, there exist $(f_{n1}),\ldots,(f_{nk})$ finite i.i.d. sequences in L_2 so that

(a) $f_{ij} \geq 0$,

(b) $\int_0^1 f_{ij}(x)dx = 1$,

(c) each (f_{nj}) has a δ-divergent zone $[a_j,b_j]$ for $1 \leq j \leq k$ so that the intervals $[a_1,b_1],\ldots,[a_k,b_k]$ are pairwise disjoint.

Proof. To prove (1) let m be a positive integer such that $1/m < \epsilon$ and let $b = 1/m$. Let $f = (1/\lambda(E))\, 1_E$ where E is a measurable set so that $\|f\| = \lambda(E)^{1-p} < \delta/m$. If $\alpha_1,\ldots,\alpha_n \geq 0$ and $\sum_{i=1}^{n} \alpha_i \leq 1$, then $|\{i : \alpha_i > b\}| < m$. Hence $\|\sum \{\alpha_i f_i : \alpha_i > b\}\| < \delta$. Now choose $0 < a < b$ so that $(a \operatorname{Var} f_1)^{p/2} < \delta$. By (9.2) $[a,b]$ is a δ-divergent zone for (f_n). (2) follows immediately from (1). □

THEOREM 9.6. If $0 \leq p < 1$, then L_p is a needle point space and, in particular, L_p contains compact convex sets with no extreme points.

Proof. By previous remarks we may concentrate on showing that the constant function 1 is a needle point in L_p, $0 < p < 1$. Let $\epsilon > 0$ and choose a positive integer k so that $(1/k)^p < \epsilon/3$. Also let $\delta = \epsilon/3k$ and choose i.i.d. sequences $(f_{n1}),\ldots,(f_{nk})$ as in Lemma 9.5 with disjoint δ-divergent zones $[a_1,b_1],\ldots,[a_k,b_k]$. By (9.2)

$$\|(1/n)\sum_{i=1}^{n} f_{ij} - 1\| < (1/n \operatorname{Var} f_{ij})^{p/2}$$

for each j, $1 \leq j \leq k$. Hence if n is chosen suitably large

$$(9.3) \qquad \|(1/n)\sum_{k=1}^{n} f_{ij} - 1\| < \epsilon/3k \quad \text{for} \quad j = 1,2,\ldots,k.$$

Let $f_i = (1/k)(f_{i1} + \ldots + f_{ik})$. We show that $F = \{f_1,\ldots,f_n\}$ satisfies conditions (1'), (2) and (3). By (9.3) $\|(1/n)(f_1+\ldots+f_n) - 1\| < \epsilon/3$ so that (1') holds. Since $\|f_{ij}\| < \delta$, $\|f_i\| < k\delta = \epsilon/3$ so that (2) holds. Now let

$\alpha_1,\ldots,\alpha_n \geq 0$ with $\Sigma\alpha_i \leq 1$. For each $j = 1,2,\ldots,k$ let

$$L_j = \Sigma\{\alpha_i f_{ij} : \alpha_i < a_j\}$$

$$M_j = \Sigma\{\alpha_i f_{ij} : \alpha_i \in [a_j,b_j]\}$$

$$R_j = \Sigma\{\alpha_i f_{ij} : \alpha_i > b_j\}.$$

Thus $\sum_{i=1}^{n} \alpha_i f_{ij} = L_j + M_j + R_j$ and

$\sum_{i=1}^{n} \alpha_i f_i = (1/k)\sum_{j=1}^{k}(L_j+M_j+R_j)$. Now $\|R_j\| < \delta$ so that

$$(9.4) \qquad \|(1/k)\sum_{j=1}^{k} R_j\| < k\delta = \epsilon/3.$$

Since the intervals $[a_1,b_1],\ldots,[a_k,b_k]$ are pairwise disjoint

$$\sum_{j=1}^{k} M_j = \Sigma\{\alpha_i f_{ij} : \alpha_i \in [a_j,b_j]\}$$

is a subconvex combination of the functions f_{ij}. Since

$\|f_{ij}\|_1 = 1$, $\|\sum_{j=1}^{k} M_j\|_1 \leq 1$. Hence

$$(9.5) \qquad \|(1/k) \sum_{j=1}^{k} M_j\| \leq (1/k)^p < \epsilon/3.$$

Finally, let

$$c_j = \Sigma\{\alpha_i : \alpha_i < a_j\}.$$

For $j = 1,2,\ldots,k$

$$\|L_j - c_j\| = \|\Sigma\{\alpha_i(f_{ij}-1) : \alpha_i < a_j\}\| < \delta.$$

Hence if $c = 1/k \sum_{j=1}^{k} c_j$, then

$$(9.6) \qquad \|((1/k) \sum_{j=1}^{k} L_j) - c\| < \delta k = \epsilon/3.$$

Since each $c_j \in [0,1]$, $c \in [0,1]$. Combining (9.4), (9.5) and (9.6) we obtain

$$\| \sum_{i=1}^{n} \alpha_i f_i - c\| < \epsilon.$$

Hence 1 is a needle point. □

REMARKS. It is possible, using the same proof, to generalize Theorem 9.6 to include a certain class of Orlicz spaces L_φ. In particular, the same proof will go through if the Orlicz function φ is chosen so that

(1) $\underline{\lim}_{x\to\infty} (\varphi(x)/x) = 0$

(2) $\overline{\lim}_{x\to\infty} (\varphi(x)/x) < \infty$.

Condition (1) is obviously necessary since L_φ must have trivial dual. It is used to select functions of the form $1/\lambda(E)\, 1_E$ with small norm in L_φ. Condition (2) is equivalent to saying that $L_1 \subset L_\varphi$, and the inclusion map is continuous (inequality (9.1) is the quantitative form of this fact for L_p, $0 < p < 1$). Also the spaces L_φ may not be transitive; however, L_∞ is dense in L_φ and multiplication by functions in L_∞ is continuous. So one still only needs to show that 1 is a needle point.

THEOREM 9.7. Let $0 < p < 1$. There exists a singular inner function S_μ such that $H^p/S_\mu H^p$ is a needle point space.

<u>Proof</u>. First observe that $\bigcup_{n=1}^{\infty} z^{-n}H^p$ is dense in L_p. Thus if $\epsilon > 0$, there exists an ϵ-needle set $F = \{f_1,\ldots,f_m\}$ for the constant function 1 with $F \subset z^{-n}H^p$ (n must be chosen suitably large). Thus $\{z^n f_1,\ldots,z^n f_m\}$ is an ϵ-needle set for z^n in H^p. As a consequence, there exists a sequence $\epsilon_n \to 0$ such that z^n possesses an ϵ_n-needle set. Now select $\beta_n \to \infty$ such that $\beta_n\epsilon_n \to 0$, let

$$\delta_n = (1/\beta_n)^{1/\gamma_2} \quad \text{and} \quad r_n = (\delta_n)^{1/n}$$

where γ_2 is the constant from the Corona Theorem in the case $n = 2$ (Theorem 3.12). Since $\delta_n \to 0$, there exists by Lemma 3.14 a singular inner function S_μ such that

$$\inf_{0 \leq \theta \leq 2\pi} |S_\mu(r_n e^{i\theta})| \geq \delta_n$$

for infinitely many n. For any such n, $|S_\mu(z)| \geq \delta_n$ if $|z| \leq r_n$ and $|z^n| \geq r_n{}^n = \delta_n$ if $|z| \geq r_n$. Thus

$$|S_\mu(z)| + |z^n| \geq \delta_n \quad \text{for every } z, \ |z| < 1.$$

By the Corona Theorem, there exist $f, g \in H^\infty$ such that

$$fS_\mu + gz^n = 1$$

with $\|f\|_\infty, \|g\|_\infty \leq (1/\delta_n)^{\gamma_2} = \beta_n$. Now suppose $h \in H^\infty$ with $\|h\|_\infty \leq 1$. Then

$$hfS_\mu + hgz^n = h.$$

Let $F = \{f_1, \ldots, f_m\}$ be a ϵ_n-needle set for z^n in H^p and let π denote the quotient map $\pi : H^p \to H^p/S_\mu H^p$. At last, $F' = \{hgf_1, \ldots, hgf_m\}$ is a $\beta_n\epsilon_n$-needle set for hgz^n (since $\|hgz^n\|_\infty \leq \beta_n$). Thus $\pi(F')$ is a $\beta_n\epsilon_n$-needle set for $\pi(hgz^n) = \pi(h - hfS_\mu) = \pi(h)$. Hence for any $h \in H^\infty$ with $\|h\|_\infty \leq 1$, $\pi(h)$ possesses a $\beta_n\epsilon_n$-needle set for infinitely many n. Since $\beta_n\epsilon_n \to 0$, $\pi(h)$ is a needle point. Thus every point in $\pi(H^\infty)$ is a needle point and since $\pi(H^\infty)$ is dense in $H^p/S_\mu H^p$, we conclude that $H^p/S_\mu H^p$ is a needle point space. □

4. Open Questions

While the examples presented in this chapter provide some insight into the nature of compact convex sets in nonlocally convex spaces, the study of this subject is far from complete and certainly is ripe for further investigation. We shall present here a list of some of the questions that remain open. The most important of these is the question posed by Klee and Schauder:

*PROBLEM 9.4: Does every compact convex set have the fixed point property?

We now define a notion which will be quite useful.

DEFINITION. An infinite dimensional compact convex subset K of an F-space is said to have the simplicial approximation property if for every $\epsilon > 0$ there exists a finite dimensional compact convex set K_0 in K such that if S is any finite dimensional simplex in K then there exists a continuous map $\psi : S \to K_0$ with

$$\|\psi(x)-x\| < \epsilon \quad \text{for} \quad x \in S.$$

Notice that the map ψ can always be chosen to be a simplicial map so that the definition really states that the identity map can be approximated by simplicial maps (into K_0). We now state without proof some equivalent versions of the simplicial approximation property in the following:

THEOREM 9.8. If K is an infinite dimensional compact convex set in an F-space then the following are equivalent:

(1) K has the simplicial approximation property

(2) For every $\epsilon > 0$, there exists $\delta > 0$ such that if K_0 is any finite dimensional compact convex subset of K with $D(K_0;K) < \delta$, then for any simplex S in K there exists a continuous map $\gamma : S \to K_0$ such that for $x \in S$, $\|\gamma(x)-x\| < \epsilon$

(3) There exists a sequence (S_n) of simplices and continuous maps $\gamma_n : S_{n+1} \to S_n$ so that

(i) $S_1 \subset S_2 \subset \ldots$

(ii) $K = \overline{\cup S_n}$

(iii) If $x \in S_{n+1}$, then $\|\gamma_n(x)-x\| < 2^{-n}$

(4) If $\epsilon > 0$, there exists a simplex S in K and a continuous map $\gamma : K \to S$ such that for every $x \in K$ $\|\gamma(x)-x\| < \epsilon$

(5) If $\epsilon > 0$, there exists $\delta > 0$ such that if K_0 is a finite dimensional compact convex set in K with $D(K_0;K) < \delta$, then there exists a continuous map $\gamma : K \to K_0$ such that for every $x \in K$ $\|\gamma(x)-x\| < \epsilon$.

REMARKS: (1) In condition (5) one can also insist that γ is a retraction onto K_0, i.e. if $x \in K_0$, then $\gamma(x) = x$.

(2) Clearly if a compact convex set has the simplicial approximation property, then it also has the fixed point property.

PROBLEM 9.5. Does every infinite dimensional compact convex set in an F-space have the simplicial approximation property?

Every nonlocally convex compact convex set that has been constructed up to the present has used needle points in the same way and it can be shown that all such sets, at least constructed so far, have the simplicial approximation property. Thus it appears that something entirely new and different is required if there is a counterexample to the Klee-Schauder question.

O. H. Keller [1931] proved that every infinite dimensional compact convex set in Hilbert space is homeomorphic to the Hilbert cube ($[0,1])^{\omega}$). It is easy to show that any metrizable compact convex set in a locally convex space can be affinely embedded in Hilbert space (use a sequence of continuous linear functionals that separate points in the set).

*PROBLEM 9.6. Is every infinite dimensional compact convex set in an F-space homeomorphic to the Hilbert cube?

It can be shown that if a compact convex set is homeomorphic to the Hilbert cube then it has the simplicial approximation property. A possible scheme for constructing counterexamples (if there are any) is as follows: find a compact convex set K with finite dimensional compact convex subsets K_n and continuous maps $T_n : K_n \to K$ so that

(1) $D(K_n;K) \to 0$

(2) if $x \in K_n$, then $\|T(x)-x\| \geq 1$.

If the sequence (T_n) is equicontinuous, then by a compactness argument one can obtain a continuous map $T : K \to K$ with no fixed point. Even if the sequence (T_n) is not equicontinuous, K will not have the simplicial approximation property (and will thus not be homeomorphic to the Hilbert cube). To see this suppose $\psi : K \to K_n$ is continuous. Then $\psi \circ T_n : K_n \to K_n$ has a fixed point $x \in K_n$ by the Brouwer Fixed Point Theorem.

Thus

$$\|\psi(T_n(x))-T_n(x)\| = \|x-T_n(x)\| \geq 1.$$

Hence if $\psi : K \to K_n$ is continuous there is always a point $y \in K$ so that $\|\psi(y)-y\| \geq 1$ and from this it follows that K does not possess the simplicial approximation property.

Another significant question with a surprising number of ramifications is the following

*PROBLEM 9.7. For which separable Banach spaces X with unit ball B does there exist a compact operator T from X into an F-space Y so that $T(B)$ is not locally convex?

An examination of the construction of the compact convex set K_0 with no extreme points shows that there is a compact operator T on L_1 with $T(B) = K_0$, i.e. L_1 is such a space. However, Kalton and Roberts [to appear] implicitly show that any compact operator on a $C(X)$-space maps the unit ball to a locally convex set. This is related to the fact that c_0 is a K-space, and will be discussed further in Chapter 10.

Another line of questions is suggested by a problem posed by J. H. Shapiro in 1976. He asked whether every F-space with trivial dual contains a compact convex set with no extreme points. Kalton [1980] answered this question by showing that in certain Orlicz spaces with trivial dual every compact convex set is locally convex. This fact suggests the following question.

*PROBLEM 9.8. Which Orlicz spaces contain compact convex sets that are not locally convex or which fail to have extreme points?

*PROBLEM 9.9. For which singular inner functions S_μ with $S_\mu H^p$ weakly dense in H^p $(0 < p < 1)$ does the quotient space $H^p/S_\mu H^p$ contain compact convex sets that are not locally convex or which fail to have extreme points?

CHAPTER 10
NOTES ON OTHER DIRECTIONS OF RESEARCH

The aim of this final chapter is to discuss rather briefly some other related topics we have not covered in the main body of the book. Our list is by no means intended to be complete and we therefore refer the interested reader to Rolewicz [1972], Waelbroeck ([1971] or [1973]) and Turpin [1976].

1. Vector measures.

There is an elegant and highly developed theory of Banach space-valued measures (see Diestel-Uhl [1976]); this theory hinges critically on local convexity, since the existence of continuous linear functionals in abundance frequently provides a reduction to the scalar case. However, attempts to extend the theory have generated some interesting problems and results.

Let X be an F-space and let Σ be some σ-algebra of sets. Suppose $\mu : \Sigma \to X$ is a σ-additive vector measure. To obtain a reasonable integration theory one must require that not only the range of μ, $\mu(\Sigma)$, is a bounded set, but also that its convex hull $\text{co}\,\mu(\Sigma)$ is bounded. Two important examples showed that this is not generally the case. First, Rolewicz and Ryll-Nardzewski [1967] showed that $\text{co}\,\mu(\Sigma)$ can be unbounded by producing a series $\sum x_n$ in an F-space so that for every subset A of $\mathbf{N}$, $\sum_{n\in A} x_n$ converges (unconditionally) but $\sum a_n x_n$ diverges for some bounded sequence (a_n). Later Turpin [1975] showed that there are measures μ so that $\mu(\Sigma)$ is already unbounded.

Fortunately, however, many special spaces do not allow this type of pathology. If X is locally bounded, for example, then $\text{co}\ \mu(\Sigma)$ is always a bounded set [Robertson 1969]. Fischer and Scholer showed this holds also for Orlicz spaces L_0 with φ unbounded [1976]. The case of L_o turned out to be very interesting. Maurey and Pisier [1973] and Kashin [1973] proved (essentially) that if $\mu(\Sigma)$ is bounded then $\text{co}\ \mu(\Sigma)$ is bounded (see also [Musial et al., 1974]). The gap was later filled by Talagrand [1981] (cf. also Kalton-Peck-Roberts [1982]) who showed that $\mu(\Sigma)$ is always bounded. Thus the theory of vector measures in L_0 behaves very well; this has some importance for the theory of stochastic integration.

Another important problem which arises very naturally is the attempt to generalize the Bartle-Dunford-Schwartz theory on existence of control measures. The problem is to determine for arbitrary $\mu : \Sigma \to X$ whether there exists a "control" measure $\lambda : \Sigma \to \mathbf{R}$ so that $\lambda(A) = 0$ implies $\mu(A) = 0$. This turns out to be nothing other than a reformulation of a classical problem of Maharam [1947] on the existence of a control measure for a continuous submeasure; this problem has been studied extensively by Christensen [Christensen-Herer, 1975], [Christensen, 1978] and Talagrand ([1979],[1980]) but remains unsolved in general. Again, for certain spaces X the problem can be solved. Talagrand [1981] shows that if $X = L_0$ then μ always has a control measure (and may effectively be treated as an L_2-valued measure). Kalton and Roberts [to appear] show that if X is locally bounded and ℓ_∞ is not finitely representable in X then again μ has a control measure. In a similar vein any measure with relatively compact range has a control measure [Kalton, Roberts, to appear]. See also [Kalton-Roberts, 1983].

In the early seventies there was considerable attention paid to attempts to generalize the Orlicz-Pettis theorem which states that, in a Banach space, a weakly countably additive vector measure is norm countably additive (see Diestel-Uhl [1976] for a survey). In [Kalton, 1971], Kalton showed that the Orlicz-Pettis theorem holds as stated if X is a separable F-space with a separating dual, and generalized the result by replacing the weak topology by any other Hausdorff vector topology weaker than the metric topology. Subsequently these results were extended and improved by a number of authors (Drewnowski [1973], [1975], Labuda [1979], Graves [1979], Pachl [1979]). See Kalton [1980] and Turpin [1978] for current related outstanding problems.

2. <u>Operators on spaces of continuous functions</u>.

Let K be a compact Hausdorff space and let $T : C(K) \to X$ be a linear operator. We shall say that T is <u>exhaustive</u> if whenever (f_n) is a uniformly bounded sequence in $C(K)$ with disjoint supports then $Tf_n \to 0$. In an important unpublished paper, Thomas [1972] showed that T is exhaustive if and only if there is a regular X-valued measure μ on the Borel sets $\mathcal{B}$ of K so that $\mathrm{co}\,\mu(\mathcal{B})$ is bounded and

$$Tf = \int f d\mu \qquad f \in C(K).$$

This extended known results on Banach spaces but required a new duality-free proof.

In Kalton [1975], Kalton extended a result of Pelczynski by showing that T is exhaustive if and only if there is no closed subspace of $C(K)$ isomorphic to c_0, on which T is an isomorphism. Later Drewnowski [1975] gave a simpler proof of the result. In the case $K = \beta N$, (so that $C(K) \approx \ell_\infty$), Drewnowski extended work of Rosenthal [1970] to show that T is exhaustive if and only if T does not preserve a copy of ℓ_∞.

The proof given by Drewnowski is very neat and elegant, and it is quite surprising that this result can be proved in a duality-free setting.

3. Tensor products

The question of topologizing the tensor product of two non-locally convex F-spaces was raised originally by Waelbroeck [1970]. He asked whether, if X and Y are F-spaces, there is a Hausdorff vector topology on the tensor product $X \otimes Y$ so that the natural bilinear form $(x,y) \to x \otimes y$ is continuous. Such a topology is called admissible. Recently Turpin ([1980] [1982a,b]) has settled this problem affirmatively.

If X and Y are p-Banach spaces, however, $X \otimes Y$ need not admit any admissible locally p-convex topology when $p < 1$; a counterexample is given by Kalton [1982]. An interesting open problem is the following: given a p-Banach space X and a q-Banach space Y, to determine for which r, $X \otimes Y$ can be given an admissible locally r-convex topology.

4. The approximation problem

If K is a compact Hausdorff space then $C(K) \otimes X$ can be identified with the subspace of $C(K,X)$ (continuous X-valued functions on K) consisting of functions with finite-dimensional range. An interesting problem raised by Waelbroeck [1972,1973] is to determine whether $C(K) \otimes X$ is dense in $C(K,X)$. This has been studied by Schuchat [1972] who gave positive results under restrictions on either K or X.

5. Algebras

Much of the theory of Banach algebras can be carried over to p-Banach algebras where $0 < p < 1$ (see Zelazko [1962]). For example if A is a commutative p-Banach algebra with identity then A admits non-trivial continuous multiplicative linear functions; in particular $A^* \neq \{0\}$. An intriguing question is whether one can have a noncommutative p-Banach algebra with identity and with trivial dual. This question arises in the context of determining whether L_p $(0 < p < 1)$ is a prime space, i.e. if every closed complemented subspace of L_p is isomorphic to L_p or is trivial. In fact (Kalton [1981]) a complemented subspace Z of L_p with $Z \not\approx L_p$ has the property that its algebra of linear endomorphisms (Z) has trivial dual.

In general the theory of "F-algebras" is less satisfactory (see [Waelbroeck 1971], [Zelazko, 1965]). An important example of Waelbroeck shows that a complex F-algebra can be a field without being isomorphic to the complex numbers.

6. Galbs

In a series of articles (e.g. [1973a], [1973b]) and a monograph [1976] Turpin introduced and studied a fundamental generalized convexity concept applicable to nonlocally convex F-spaces. If X is an F-space the galb o X, G(X), is defined to be the space of all sequences $(a_n)_{n=1}^{\infty}$ so that whenever (x_n) is a bounded sequence in X then the series $\sum a_n x_n$ converges. Thus X is locally convex if and only if $G(X) = \ell_1$, and in general $G(X) \subset \ell_1$. Associated wit the space G(X) is a convergence structure on G(X) (either a topology or a bornology). Turpin used the notion of a galb effectively in the study of linear operators between certain types of Orlicz function spaces.

Certain galb conditions appear to be of special interest. In [Kalton, 1979], an F-space X is said to be strictly galbed if there is a sequence $(a_n) \in G(X)$ with $a_n > 0$ for every n; this eliminates from consideration certain spaces such as L_0 for which G(X) is essentially trivial. I is shown in [Kalton, 1979] that a non-locally bounded strictly galbed space contains an infinite-dimensional locally convex subspace.

If X is locally bounded then $G(X) \supset \ell_p$ for some $p > 0$; this is essentially the Aoki-Rolewicz theorem proved in Chapter 1. More generally Turpin [Turpin, 1976] showed that the class of <u>exponentially galbed</u> spaces, i.e. $(2^{-n})_{n=1}^{\infty} \in G(X)$, are well-behaved for many applications.

In [Kalton, 1981] Kalton studied the galbs of certain twisted sums. A quasi-Banach space X is said to be <u>log convex</u> if its galb contains all sequences (a_n) so that $\sum |a_n| < \infty$ and $\sum |a_n| \, |\log (1/|a_n|)| < \infty$. It is shown that the twisted sum of two Banach spaces is always log convex, and conversely any log convex quasi-Banach space is isomorphic to the quotient of a subspace of a twisted sum of two copies of ℓ_1.

LIST OF REFERENCES

Aleksandrov, A. B.; Invariant subpsaces of the backward shift operator in the space H^p, $p \in (0,1)$ (in Russian), Zap. Naucn. Sem. Leningrad. Otdel. Mat. Inst. Steklov (LOMI) 92(1979), 7-29, 318.

_________________; Approximation by rational functions and an analogue of M. Riesz's theorem on conjugate functions for the space L^p with $p \in (0,1)$ (in Russian), Mat. Sbornik (N.S.) 107(149) 1978, no. 1, 3-19, 159.

_________________; Essays on the nonlocally convex Hardy class, in Complex analysis and spectral theory, Springer Lecture Notes 864, Berlin-Heidelberg-New York, 1981.

Anderson, R. D.; Hilbert space is homeomorphic to the countable infinite product of lines, Bull. Amer. Math. Soc. 72(1966), 515-519.

Aoki, T.; Locally bounded linear topological spaces, Proc. Imp. Acad. Tokyo 18(1942), No. 10.

Banach, S.; Theorie des operations lineaires, Monografie Matematyczne I, Waresaw 1932.

Beck, A.; A convexity condition in Banach spaces and the strong law of large numbers, Proc. Amer. Math. Soc. 13(1962), 329-334.

Berg, I.D., Peck, N.T., and Porta, H.; There is no continuous projection from the measurable functions on the square onto the measurable functions on the interval, Israel J. Math. 14(1973), 199-204.

Bergh, J., and Lofstrom, J.; Interpolation spaces, an introduction, Grundlehr. der math. Wissensch. 223, Springer, Berlin-Heidelberg-New York, 1976.

Bessaga, C.; On topological classification of complete linear metric spaces, Fund. Math. 56(1965), 251-288.

Bessaga, C., and Pelczynski, A.; Some remarks on homeomorphisms of F-spaces, Bull. Acad. Polon. Sci. 10(1962), 265-270.

__________________________; The space of Lebesgue measurable functions on the interval [0,1] is homeomorphic to the countable infinite product of lines, Math. Scand. 27(1970) 132-140.

Beurling, A.; On two problems concerning linear transformations in Hilbert spaces, Acta Math. 81(1949), 239-255.

Bourgain, J.; A counterexample to a complementation problem, Compositio Math. 43(1981), 133-144.

Bourgain, J. and Rosenthal, H.P.; Martingales valued in certain subspaces of L_1, Israel J. Math. 37(1980), 54-75.

Bourgin, D. G.; Linear topological spaces, Amer. J. Math 65(1943), 637-659.

Carleson, L.; An explicit unconditional basis in H_1, Bull. Sci. Math. 104(1980), 405-416.

Cater, S.; Continuous linear functionals on certain topological vector spaces, Pacific J. Math. 13(1963), 65-71.

Christensen, J.P.R.; Some results with relation to the control measure problem, pp. 27-34 in Vector measures and applications II, Springer Lecture Notes 645, Berlin-Heidelberg-New York, 1978.

Christensen, J.P.R., and Herer, W.; On the existence of pathological submeasures and the construction of exotic groups, Math. Ann. 213(1975), 203-210.

Corson, H. and Klee, V.L.; Topological classification of convex sets, Proc. Symposium on Convexity, Proc. Symp. Pure Math. No. 7, Amer. Math. Soc. 1963, 37-51.

Day, M.M.; The spaces L_p with $0 < p < 1$, Bull. Amer. Math. Soc. 46(1940), 816-823.

Dierolf, S.; Uber Vererbbarkeitseigenschaften in topologischen Vektorraumen, Dissertation, Munich 1974.

Diestel, J. and Uhl, J.J.; Vector Measures, American Math. Soc. Surveys, Providence, 1976.

Drewnowski, L.; On the Orlicz-Pettis type theorems of Kalton, Bull. Acad. Polon. Sci. 21(1973), 515-518.

______________; Another note on Kalton's theorems, Studia Math. 52(1975), 233-237.

______________; On minimally subspace-comparable F-spaces, J. Functional Analysis 26(1977a), 315-332.

______________; The weak basis theorem fails in non-locally convex F-spaces, Canad. J. Math 29(1977b), 1069-1071.

______________; On minimal topological linear spaces and strictly singular operators, Comment. Math. (Ser. Spec.) II(1979), 89-106.

Duren, P. L.; Theory of H^p-spaces, Academic Press, New York-London, 1970.

Duren, P. L., Romberg, B.W., and Shields, A.L.; Linear functionals on H^p spaces when $0 < p < 1$, J. Reine Angew. Math 238(1969), 32-60.

Enflo, P.; A counterexample to the approximation problem in Banach spaces, Acta Math. 130(1973), 309-317.

Enflo, P., Lindenstrauss, J., and Pisier, G.; On the "three space problem", Math. Scand. 36(1975), 199-210.

Enflo, P., and Starbird, T.W.; Subspaces of L_1 containing L_1, Studia Math. 65(1979), 203-225.

Fischer, W. and Scholer, U.; The range of vector measures into Orlicz spaces, Studia Math. 59(1976), 53-61.

Gamelin, T.W.; H^p spaces and extremal functions in H^1, Trans. Amer. Math. Soc. 124(1966), 158-167.

______________; Wolff's proof of the Corona theorem, Israel J. Math. 37(1980), 113-119.

Garling, D.J.H.; Sums of Banach space valued random variables, Lecture notes, Ohio State University, 1977.

Giesy, D.P.; On a convexity condition in normed linear spaces, Trans. Amer. Math. Soc. 125(1966), 114-146.

Graves, W.H.; Universal Lusin measurability and subfamily summable families in abelian topological groups, Proc. Amer. Math. Soc. 235(1979), 45-50.

Gregory, D.A. and Shapiro, J.H.; Nonconvex linear topologies with the Hahn-Banach extension property, Proc. Amer. Math. Soc. 25(1970), 902-905.

Halmos, P.; Measure Theory, Van Nostrand, Princeton,1950.

Hyers, D.H.; Locally bounded linear topological spaces, Revista Ci. 41(1939), 555-574.

Jamison, R.E., O'Brien, R.C., and Taylor, P.D.; On embedding a compact convex set into a locally convex topological vector space, Pacific J. Math. 64(1976), 129-130.

Kadec, M.I.; Proof of the topological equivalence of all separable infinite-dimensional Banach spaces, Translated from Funktsional'nyi Analiz i. Ego Prilozheniya, 1, 1, 61-70, submitted October 9, 1966.

Kalton, N.J.; Subseries convergence in topological groups and vector spaces, Israel J. Math. 10(1971), 402-412.

____________; Basic sequences in F-spaces and their applications, Proc. Edinburgh Math. Soc. (2) 19(1974), 151-167.

____________; Exhaustive operators and vector measures, Proc. Edinburgh Math. Soc. 19(1975), 291-300.

____________; Compact and strictly singular operators on Orlicz spaces, Israel J. Math. 26(1977a), 126-136.

____________; Universal spaces and universal bases in metric linear spaces, Studia Math. 61(1977b), 161-191.

____________; Linear operators whose domain is locally convex, Proc. Edinburgh Math. Soc. 20(1977c), 293-300.

____________; The endomorphisms of L_p, $0 \leq p \leq 1$, Indiana Univ. Math. J. 27(1978a), 353-381.

____________; The three space problem for locally bounded F-spaces, Comp. Math. 37(1978b), 243-276.

____________; Transitivity and quotients of Orlicz spaces, Comment Math. Special Issue 1(1978c), 159-172.

____________; Compact operators on symmetric function spaces, Bull. Acad. Polon. Sci. 26(1978d), 815-816.

____________; Quotients of F-spaces, Glasgow Math. J. 19(1978e), 103-108.

____________; A note on galbed spaces, Comm. Math. (1979), 75-79.

____________; Linear operators on L_p for $0 < p < 1$, Trans. Amer. Math. Soc. 259(1980a), 319-355.

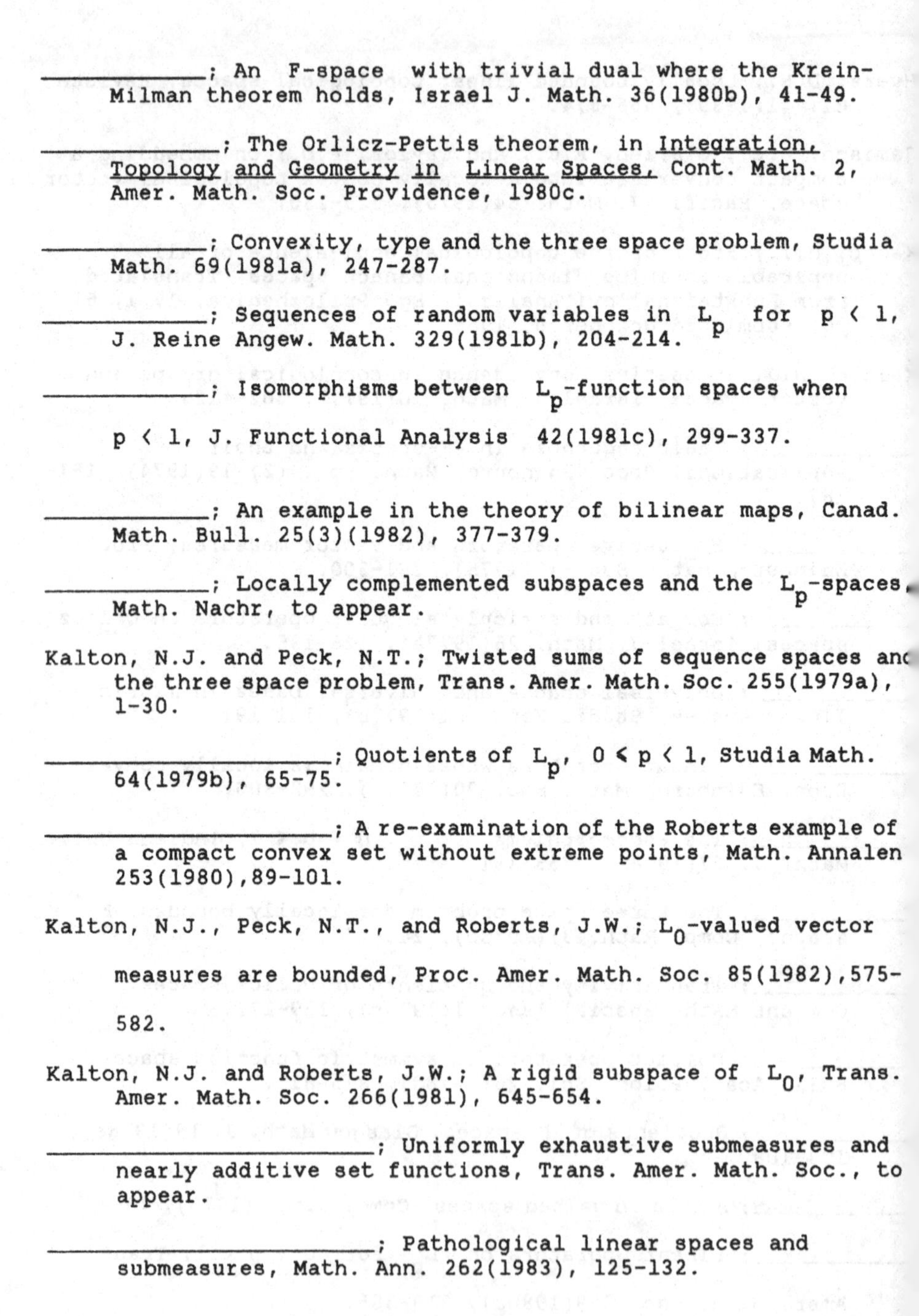

______; An F-space with trivial dual where the Krein-Milman theorem holds, Israel J. Math. 36(1980b), 41-49.

______; The Orlicz-Pettis theorem, in <u>Integration, Topology and Geometry in Linear Spaces,</u> Cont. Math. 2, Amer. Math. Soc. Providence, 1980c.

______; Convexity, type and the three space problem, Studia Math. 69(1981a), 247-287.

______; Sequences of random variables in L_p for $p < 1$, J. Reine Angew. Math. 329(1981b), 204-214.

______; Isomorphisms between L_p-function spaces when $p < 1$, J. Functional Analysis 42(1981c), 299-337.

______; An example in the theory of bilinear maps, Canad. Math. Bull. 25(3)(1982), 377-379.

______; Locally complemented subspaces and the L_p-spaces, Math. Nachr, to appear.

Kalton, N.J. and Peck, N.T.; Twisted sums of sequence spaces and the three space problem, Trans. Amer. Math. Soc. 255(1979a), 1-30.

______; Quotients of L_p, $0 \leq p < 1$, Studia Math. 64(1979b), 65-75.

______; A re-examination of the Roberts example of a compact convex set without extreme points, Math. Annalen 253(1980),89-101.

Kalton, N.J., Peck, N.T., and Roberts, J.W.; L_0-valued vector measures are bounded, Proc. Amer. Math. Soc. 85(1982),575-582.

Kalton, N.J. and Roberts, J.W.; A rigid subspace of L_0, Trans. Amer. Math. Soc. 266(1981), 645-654.

______; Uniformly exhaustive submeasures and nearly additive set functions, Trans. Amer. Math. Soc., to appear.

______; Pathological linear spaces and submeasures, Math. Ann. 262(1983), 125-132.

Kalton, N.J. and Shapiro; J.H.; An F-space with trivial dual and non-trivial compact endomorphisms, Israel J. Math. 20(1975), 282-291.

__________________________; Bases and basic sequences in F-spaces, Studia Math. 56(1976), 47-61.

Kalton, N.J. and Trautman, D.A.; Remarks on subspaces of H_p where $0 < p < 1$, Michigan Math. J. 29(1982), 163-176.

Kanter, M.; Stable laws and embedding of L_p-spaces, Amer. Math. Monthly 80(1973), 403-407.

Kashin, B.S.; The stability of unconditional almost everywhere convergence (in Russian) Mat. Zametki 14(1973),645-654.

Kelley, J.L.; General Topology, Van Nostrand, New York, 1955.

Kelley, J.L., Namioka, I., and co-authors, Linear topological spaces, Van Nostrand, Princeton, 1963.

Keller, O.-H.; Die Homomorphie der kompakten konvexen Mengen im Hilbertschen Raum, Math. Annalen 105(1931), 748-758.

Klee, V.L.; Invariant metrics in groups, Proc. Amer. Math. Soc. 3(1952), 484-487.

__________; Convex bodies and periodic homeomorphisms in Hilbert space, Trans. Amer. Math. Soc. 74(1953), 10-43.

__________; Some topological properties of convex sets, Trans. Amer. Math. Soc. 78 (1955), 30-45.

__________; An example in the theory of topological linear spaces, Arch. Math. 7(1956), 362-366.

__________; Leray-Schauder theory without local convexity, Math. Annalen 141(1960), 286-296.

__________; Exotic topologies for linear spaces, Proc. Symposium on general topology and its relations to modern algebra, Prague, 1961a.

__________; Relative extreme points, Proc. Symp. on linear spaces, Jerusalem, 1960, Pergamon, London, 1961b, 282-289.

Koosis, P.; Introduction to H^p spaces, London Math. Soc.

Lecture note series, 40, Cambridge press, Cambridge-New York, 1980.

Korenblum, B.; Cyclic elements in some spaces of analytic functions, Bull. Amer. Math. Soc. (new series), 5(1981), 317-318.

Kothe, G.; Topological Vector Spaces I, Springer, Berlin-Heidelberg-New York, 1969.

Kwapien, S.; On the form of a linear operator in the space of all measurable functions, Bull. Acad. Polon. Sci. 21(1973), 951-954.

Labuda, I.; Universal measurability and summable families in tvs, Proc. Kon. Ned. Akad. A 82(1979), 27-34.

Landsberg, M.; Lineare topologische Raume die nicht lokalkonvex sind, Math. Z. 65(1956), 104-112.

Lawson, J.D.; Embeddings of compact convex sets and locally compact cones, Pacific J. Math. 66(1976), 443-453.

Lewis, D.R. and Stegall, C.; Banach spaces whose duals are isomorphic to $\ell_1(\Gamma)$, J. Functional Analysis 12(1973), 177-187.

Lindenstrauss, J.; On a certain subspace of ℓ_1, Bull. Acad. Polon. Sci. 12(1964), 539-542.

Lindenstrauss, J. and Pelczynski, A.; Contributions to the theory of classical Banach spaces, J. Functional Analysis 2(1971), 225-249.

Livingston, A.E.; The space H^p $0 < p < 1$ is not normable, Pacific J. Math. 3(1953), 613-616.

Maharam, D.; An algebraic characterization of measure algebras, Ann. Math. (2) 48(1947), 154-167.

W. Matuszewska and W. Orlicz; A note on the theory of s-normed spaces of φ-integrable functions, Studia Math. 21(1961), 107-115.

Maurey, B.; Nouveaux theoremes de Nikishin, Seminaire Maurey-Schartz 1973-74, no's 4,5, Ecole Polytechnique, Palaiseau, 1974.

__________; Isomorphismes entre espaces H_1, Acta Math. 145(1980), 79-120.

Maurey, B. and Pisier, G.; Un theoreme d'extrapolation et ses consequences, C.R. Acad. Sci. (Paris) A277 (1973), 39-42.

Morrow, J.R.; On basic sequences in non-locally convex spaces, Studie Math. 67(1980), 119-133.

Musial, K., Ryll-Nardzewski, C., and Woyczynski, W.A.; Convergence presque sure des series aleatoires vectorielles a multiplicateurs bornes, C.R.Acad. Sci. (Paris) Ser A 279(1974), 225-228.

Nikishin, E.M.; Resonance theorems and superlinear operators, Transl. of Uspekhi Mat. Nauk. XXV no. 6 Nov. Dec. 1970.

__________; Izvestia Akad. Nauk. S.S.S.R., 36, no. 4, 1972.

Nikodym, O.; Contribution a la theorie des fonctionnelles lineaires en connexion avec la theorie de la mesure des ensembles astraits. Mathematica. Cluj, 5(1931), 130-141.

Oswald, P.; On spline bases in periodic Hardy spaces $(0 < p < 1)$, Math. Nach. 108(1982), 219-229.

__________; On Schauder bases in Hardy spaces, Proc. Roy. Soc. Edinburgh A93(1983), 259-263.

achl, J.; A note on the Orlicz-Pettis theorem, Proc. Kon. Ned. Akad. A82(1979), 35-37.

allaschke, D.; The compact endomorphisms of the metric linear spaces L_φ, Studia Math. 47(1973), 123-133.

eck, N.T.; On non-locally convex spaces I, Math. Ann. 161(1965), 102-115.

eck, N.T. and Starbird, T.; L_0 is ω-transitive, Proc. Amer. Math. Soc. 83(1981), 700-704.

eetre, J.; Locally analytically pseudo-convex topological vector spaces, Studia Math. 73(1982), 253-262.

elczynski, A.; Projections in certain Banach spaces, Studia Math. 19(1960), 209-228.

ettis, B.J.; On continuity and openness of homomorphisms in topological groups, Ann. Math. 52(1950), 293-308.

isier, G.; Sur les espaces qui ne contiennent pas uniformement de ℓ_n^1, Seminaire Maurey-Schwartz 1973-74, Ecole Polytechnique, Paris, Expose VII.

Ribe, M.; On the separation properties of the duals of general topological vector spaces, Arkiv for Math. 9(1971), 279-302.

________; Necessary convexity conditions for the Hahn-Banach theorem in metrizable spaces, Pacific J. Math. 44(1973), 715-732.

________; Examples for the nonlocally convex three space problem, Proc. Amer. Math. Soc. 237(1979), 351-355.

Roberts, J. W.; Pathological compact convex sets in the spaces $L_p(0,1)$, $0 \leqslant p < 1$, The Altgeld Book, X, Univ. of Illinois Functional Analysis Seminar, 1975-76.

________; A compact convex set with no extreme points, Studia Math. 60(1977a), 255-266.

________; A nonlocally convex F-space with the Hahn-Banach approximation property, Banach spaces of analytic functions, Springer Lecture Notes 604, Berlin-Heidelberg-New York, 1977, 76-81.

________; The embedding of compact convex sets in locally convex spaces, Canad. J. Math. 30(1978), 449-454.

________; Cyclic inner functions in weighted Bergman spaces and weak outer functions in H_p, $0 < p < 1$, Illinois J. Math, to appear.

Robertson, A. P.; On unconditional convergence in topological vector spaces, Proc. Roy. Soc. Edinburgh A68(1969), 145-157.

Rolewicz, S.; On a certain class of linear metric spaces, Bull. Acad. Polon. Sci. 5(1957), 471-473.

________; Some remarks on the spaces N(L) and N(ℓ), Studia Math. 18(1959), 1-9.

________; Metric Linear Spaces, Monografie Matematyczne 56, PWN, Warsaw, 1972.

Rolewicz, S. and Ryll-Nardzewski, C.; On unconditional convergence in linear metric spaces, Colloq. Math. 17(1967), 327-331.

Rosenthal, H.P.; On relatively disjoint families of measures with some applications to Banach space theory, Studia Math. 37(1970), 13-31.

Royden, H.L.; Real Analysis, Macmillan, New York, 1963.

Rudin, W.; Functional Analysis, McGraw-Hill, New York, 1973.

Schaefer, H.; Banach lattices and positive operators, Springer, Berlin-Heidelberg-New York, 1974.

Schuchat, A.H.; Approximation of vector-valued continuous functions, Proc. Amer. Math. Soc. 31(1972),97-103.

Shapiro, J.H.; Examples of proper closed weakly dense subspaces in some F-spaces of analytic functions, Israel J. Math. 7(1969), 369-380.

______________; Extension of linear functionals on F-spaces with bases, Duke Math. J. 37(1970), 639-645.

______________; On the weak-basis theorem in F-spaces, Canad. J. Math. 26(1974), 1294-1300.

______________; Mackey topologies, reproducing kernels, and diagonal maps on the Hardy and Bergman spaces, Duke Math. J. 43(1976), 197-202.

______________; Remarks on F-spaces of analytic functions, Banach spaces of analytic functions, Lecture Notes in Math. 604, Springer, Berlin-Heidelberg-New York, 1977, 107-124.

imons, S.; The sequence spaces $\ell(p_\nu)$ and $m(p_\nu)$, Proc. London Math. Soc. 15(1965),422-436.

tiles, W.J.; On properties of subspaces of ℓ_p, $0 < p < 1$, Trans. Amer. Math. Soc. 149(1970), 405-415.

______________; Some properties of ℓ_p, $0 < p < 1$, Studia Math. 42(1972), 109-119.

erry, W.E.; F-spaces universal with respect to linear codimension, Proc. Amer. Math. Soc. 63(1977), 59-65.

alagrand, M.; A simple example of a pathological submeasure, Math. Ann. 252(1980), 97-102.

______________; Les mesures vectorielles a valeurs dans L_0 sont bornees, Ann. Sci. Ecole Norm. Sup. 24(1981), 445-452.

Thomas, G.E.F.; On Radon maps with values in arbitrary topological vector spaces and their integral extensions, unpublished manuscript, 1972.

Turpin, P.; Operateurs lineaires entre espaces d'Orlicz non localement convexes, Studia Math. 46(1973a), 153-163.

__________; Espaces et intersections d'espaces d'Orlicz non localement convexes, Studia Math. 46(1973b), 167-195.

__________; Une mesure vectorielle non bornee, C. R. Acad. Sci. (Paris) A 280(1975), 509-511.

__________; Convexites dans les espaces vectoriels topologiques generaux, Diss. Math. 131, Warsaw, 1976.

__________; Properties of Orlicz-Pettis or Nikodym type and barrelledness conditions, Ann. Inst. Fourier (Grenoble) 28(3)(1978), 67-86.

__________; Sur certains produits tensoriels topologiques, C. R. Acad. Sci. (Paris) 291(1980), 405-407.

__________; Representation fonctionelle des espaces vectoriels topologiques, Studia Math 63(1982a), 1-10.

__________; Produits tensoriels d'espaces vectoriels topologiques, Bull. Soc. Math. France 110(1982b), 3-14.

Waelbroeck, L.; The tensor product of a locally pseudo-convex and a nuclear space, Studia Math. 38(1970), 101-104.

______________; Topological vector spaces and algebras, Lectur notes 230, Springer, Berlin-Heidelberg-New York, 1971.

______________; Topological vector spaces, Summer school on topological vector spaces, Bruxelles 1972, Springer Lecture Notes 331, Berlin-Heidelberg-New York, 1973, 1-40.

______________; A rigid topological vector space, Studia Math. 59(1977), 227-234.

Williamson, J.H.; Compact linear operators in linear topologica spaces, J. London Math. Soc. 29(1954), 149-156.

Wiweger, A.; Linear spaces with mixed topology, Studia Math. 20(1961), 47-68.

Wojtaszczyk, P.; The Franklin system is an unconditional basis in H_1, Arkiv. for Math. 20(1982), 293-300.

________________; H_p-spaces ($p < 1$) and spline systems, to appear.

Zelazko, W.; On the locally bounded and m-convex topological algebras, Studia Math. 19(1960), 333-356.

___________; On the radicals of p-normed algebras, Studia Math. 21(1962), 203-206.

___________; Metric generalizations of Banach algebras, Diss. Math. 47, Warsaw 1965.

GLOSSARY OF TERMS

This index gives the page numbers for the definitions of terms and symbols we use. Terminology which is standard or which is defined and used in only one section may not be listed.